Realistic Applications in Mechanics

REALISTIC APPLICATIONS IN MECHANICS

by

THE SPODE GROUP

Oxford New York Tokyo
OXFORD UNIVERSITY PRESS

Oxford University Press, Walton Street, Oxford OX2 6DP
Oxford New York Toronto
Delhi Bombay Calcutta Madras Karachi
Petaling Jaya Singapore Hong Kong Tokyo
Nairobi Dar es Salaam Cape Town
Melbourne Auckland
and associated companies in
Berlin Ibadan

Oxford is a trade mark of Oxford University Press

Published in the United States
by Oxford University Press, New York

© The Spode Group, 1986

First published 1986
Reprinted 1989

Individual pages of this book may be reproduced
by individual teachers for class use
without the permission from the publisher.
With this exception, the material remains copyright
and no part of this publication may be reproduced
without the prior permission of Oxford University Press

British Library Cataloguing in Publication Data
Realistic applications in mechanics.
1. Mechanics, Applied
I. Spode Group
620.1 TA350
ISBN 0–19–853654–2

Library of Congress Cataloging in Publication Data
Main entry under title:
Realistic applications in mechanics.
1. Mechanics, Applied. I. Spode Group.
TA350.R37 1985 620.1 85–25814
ISBN 0–19–853654–2 (pbk.)

Printed in Great Britain by
St Edmundsbury Press,
Bury St Edmunds, Suffolk

Contents

List of contributors

John Berry (Editor)	Faculty of Mathematics, The Open University
Roger Biddlecombe	Ounsdale High School, Wolverhampton
Roger Blackford	Computing Advisory Teacher, Stafford
Brian Bolt	School of Education, Exeter University
Morag Borrie	Pilgrim School, Bedford
David Burghes (Editor)	Professor of Education, Exeter University
Bob Davison	School of Mathematics, Leicester Polytechnic
Bob Francis	Chester College of Higher Education
Nigel Green	Tong School, Bradford
Ron Haydock	Matlock College of Higher Education
Ian Huntley (Editor)	Department of Math Sciences, Sheffield City Polytechnic
Paula Sellwood	Maths Advisory Teacher, Hertfordshire
John Walton	Monks Walk School, Hertfordshire

ILLUSTRATIONS

Nigel Weaver	School of Education, Exeter University

THE SPODE GROUP

Director	Professor David Burghes
Associate directors	Dr John Berry and Dr Ian Huntley
Secretary	Ms Sally Williams, School of Education, University of Exeter, St. Luke's, Exeter EX1 2LU, UK

Introduction

There has been a tendency in recent years for mechanics to become rather an esoteric subject concerned with particles, smooth planes, inelastic weightless strings, and so on. This approach tends to concentrate on the fundamental postulates and the solutions of the differential equations involved. The students then lose sight of one of the aims of mechanics which is to provide a model for the motion of bodies in the 'real world'. Of course the fundamentals of mechanics do involve assumptions that make the ideal particle approach appropriate but we must not forget that one of the uses of mechanics is to investigate problems of physical or technological importance. Thus, mechanics should be an interplay between experience and experiment (the real world) on the one hand and fundamental concepts (the mathematical world) on the other.

The authors have written this book with the intention of providing a collection of applications of mechanics for the teacher or student to appreciate relevant uses of mechanics. We hope that they will provide a useful set of examples that are both convincing applications of mechanics and applications that are suitable for the beginner (both at school and in higher education). There are 17 chapters in this book covering most of the syllabus for a first course in mechanics. Each chapter consists of a statement of the problem and a detailed solution. The part of the syllabus for which the problem is appropriate is identified at the beginning of each chapter. The authors believe that the problems in this book can be used in three ways.

1. To provide *motivation* for the development of a particular part of the syllabus;
2. To provide 'real problems' to illustrate the use of new theory that has recently been taught to the students, showing possible *applications*;
3. To provide *interesting* problems different from the artificiality of the traditional sort of examination question.

We very much hope that this collection of problems in this text will be of help in making mechanics more interesting and more relevant.

THE SPODE GROUP

This is the name of the Group of authors of this text. We are a small group of mathematics teachers from schools, colleges, and higher education who have come together to provide practical material for the mathematics classroom. The material was developed at weekend working meetings of the Group, which took place at the Dominican Priory at Spode, Staffordshire. The Group has already produced three earlier publications,

Solving real problems with mathematics, Volume 1

Solving real problems with mathematics, Volume 2

Solving real problems with C.S.E. mathematics,

published by Cranfield Press.

For further details of the Group's activities, write to

Professor David N. Burghes
Director, Spode Group
University of Exeter
School of Education
St. Luke's, EX1 2LU, UK

Acknowledgements

The first draft of this book was written at a weekend meeting at Spode Conference Centre in December 1982. We are delighted to acknowledge the Warden and his staff for making our stay so welcome and for providing a suitable working environment.

We are grateful for financial assistance from the Institute of Mathematics and its Applications.

Finally we would like to express our thanks to Mr N. Weaver for providing the artwork throughout this text.

1. Sports

1.1. *N* men in a boat

Components and moments of forces

Rowing boats are traditionally designed for one, two, four, or eight oarsmen. There are several ways that oarsmen can be seated so that as much as possible of their efforts goes into moving the boat forward, rather than turning it from side to side.

Clearly, with a single oarsman in a sculling boat with two oars opposite each other, the force **F** which each oar exerts on the boat through the rowlock can be resolved into components parallel (P) and normal (N) to the direction of movement (Fig. 1.1). The normal components are equal and opposite and the parallel components add to drive the boat straight forward with force $2P$.

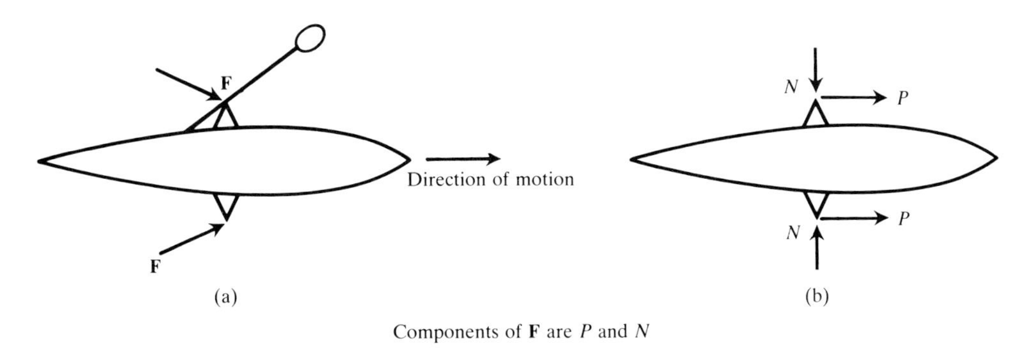

Components of **F** are P and N

Fig 1.1

Notice that once the oar has passed through the position at right angles to the boat (where N is zero and **F** is at its most effective, parallel to P) the component N reverses its direction (Fig. 1.2).

N men in a boat

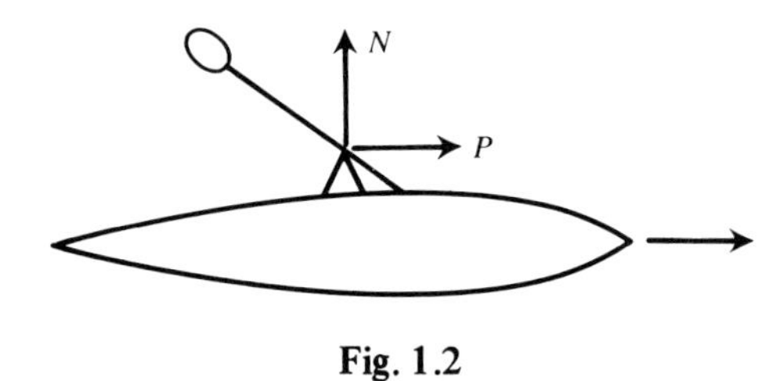

Fig. 1.2

In a pair, one oarsman must sit behind the other. Assuming both men exert the same force **F** simultaneously, the normal components will still be equal and opposite, so the effect in *translation* is to move the boat forward with a net force $2P$. Taking moments about any point on the centre line of the boat, however, shows that during the first half of the stroke the boat tends to rotate anticlockwise. For instance, if we take moments about 0 (referring to Fig. 1.3), with clockwise as positive, we have

$$\text{Net moment about } 0 = Nd + (-N)(d + s)$$
$$= -Ns,$$

i.e. a net turning effect of Ns anticlockwise about the stern. Once the oars have passed the right-angle position, each of the normal components reverses its direction and the net moment becomes Ns clockwise.

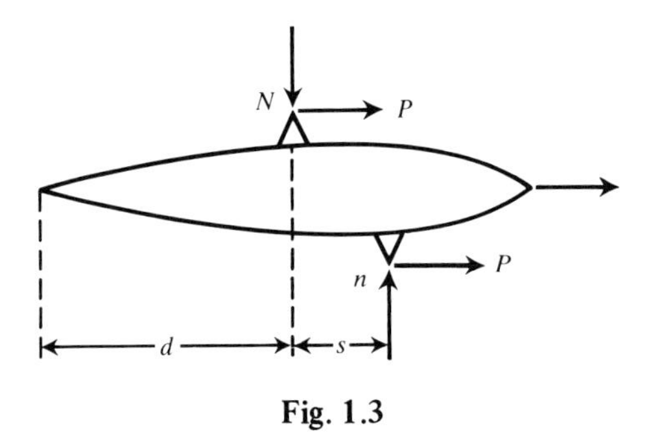

Fig. 1.3

Thus the observed effect is that the boat proceeds through the water with a characteristic wiggle, turning slightly one way, then the other as it moves forward.

Analyse in the same way the two possible arrangements of the oars in a four (Fig. 1.4).

Fig. 1.4

• Solution

Traditional four (Fig. 1.5)

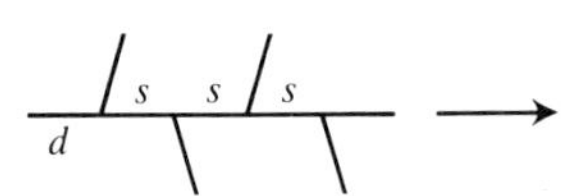

Fig. 1.5. Traditional four.

The net moment about the stern in first half of the stroke

$$= +Nd + -N(d+s) + N(d+2s) + -N(d+3s)$$
$$= -2Ns$$

and the net moment about the stern in the second half of the stroke is $+2Ns$.

Italian four (Fig. 1.6)

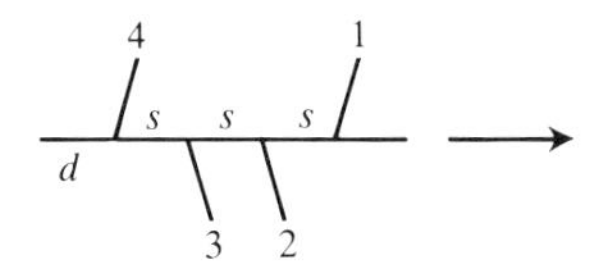

Fig. 1.6. Italian four.

The net moment in first half of the stroke

$$= +Nd + -N(d+s) + -N(d+2s) + N(d+3s)$$
$$= 0.$$

It is also zero in the second half, so that this way of 'rigging' the boat is mechanically much more efficient since there is no net moment at any time and hence no wasted effort in wiggling motion. The rig has the disadvantage however that the stroke man 4 tends to put his oar into the rough puddle caused by the bow man 1, especially as the boat gets up speed.

It is of course possible to offset the turning effect in the traditional rigs by taking account of strength differences in the oarsmen.

• Problem

For which of the following eight-man positions is the turning effect zero?

(a) Traditional eight

(b) Italian eight

(c) German eight

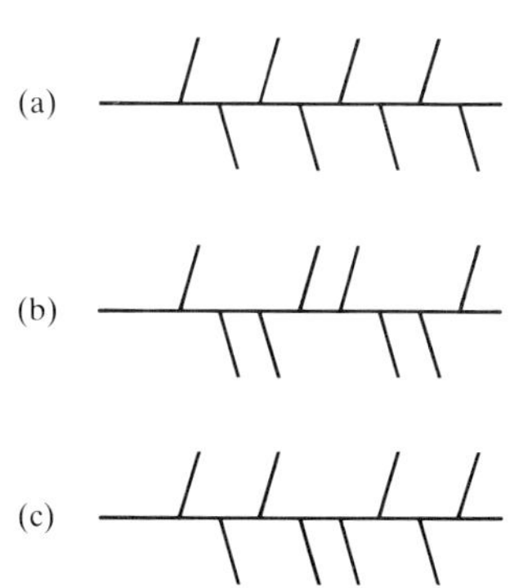

Fig. 1.7. (a) Traditional eight; (b) Italian eight; (c) German eight.

• Further reading

Townend, S. M. (1982). *J. Teaching Maths and Its Applications* 1, pp. 18–20.

1.2. High-jump techniques

Projectiles; impulse; centre of mass

Like all other field events nowadays, the high jump requires special techniques which have developed during the history of the event. Originally the most popular technique was the scissor jump (Fig. 1.8). This technique is easy to master but mechanically inefficient, since the centre of mass needs to be raised to a considerable height above the bar.

Fig. 1.8. Scissors jump.

As training methods and landing areas improved, the techniques of Eastern cut off and Western roll were developed, and from these the straddle jump emerged (Fig. 1.9). This technique is still popular with high jumpers from the Iron Curtain countries and involves the athlete wrapping himself around the bar. This is obviously more efficient, since the centre of mass only has to be raised to a level just above the bar.

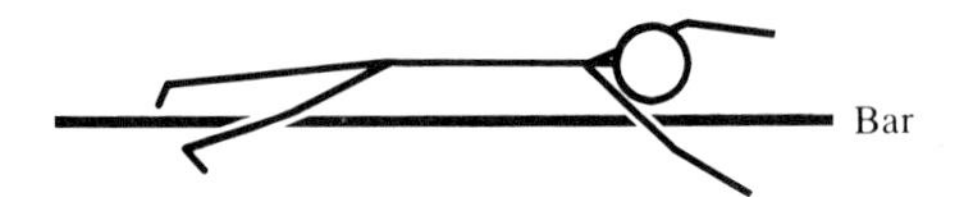

Fig. 1.9. Straddle jump.

The sport was revolutionized in 1968 by an American high jumper, Dick Fosbury, who developed the flop technique, which involves the athlete travelling over the bar with his back arched (Fig. 1.10). The Iron Curtain countries have now begun to use this technique.

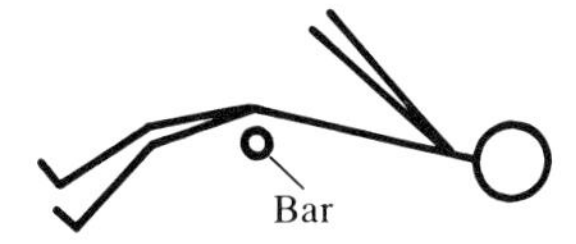

Fig. 1.10. Fosbury flop.

Compare the techniques of scissor jump, straddle, and Fosbury flop from the point of efficiency.

• Solution

Whichever technique is used, the basic problem is to convert horizontal speed into vertical lift in order to execute the jump.

Suppose the athlete of mass m, approaches the bar with a horizontal velocity u. At take-off he experiences a vertical impuse I (i.e. he jumps), and immediately after this the horizontal and vertical velocity components of his centre of mass are U and V respectively.

Using the principle of conservation of momentum we obtain

Horizontally

$$m(U-u)=0 \tag{1.2.1}$$

Vertically

$$m(V-0)=I. \tag{1.2.2}$$

Thus we have

$$U=u \quad \text{and} \quad V=I/m. \tag{1.2.3}$$

We denote the height of the centre of mass of the jumper at the moment of take-off by h. Then the flight of the jumper can be thought of as a standard projectile problem as shown in Fig. 1.11. Applying Newton's second law of motion in the horizontal direction gives

$$m\frac{\mathrm{d}^2x}{\mathrm{d}t^2}=0.$$

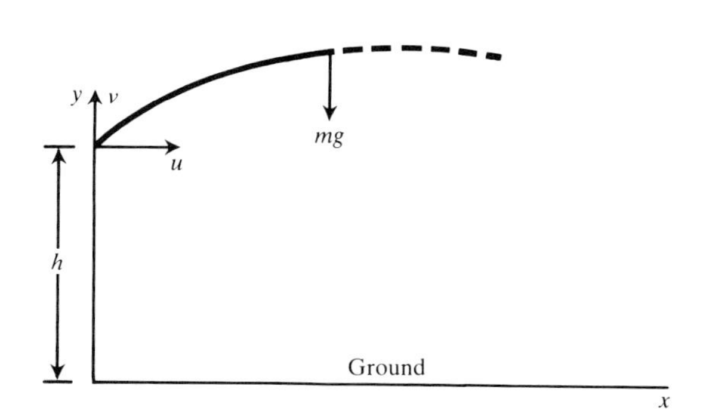

Fig. 1.11

The initial conditions are $x = 0$, $u = U$ at time $t = 0$. Integrating twice and using these initial conditions gives

$$\frac{dx}{dt} = U$$

$$x = Ut.$$

Proceeding similarly for the vertical direction gives

$$m\frac{d^2y}{dt^2} = -mg$$

with $y = h$, $v = V$ at time $t = 0$, so that

$$\frac{dy}{dt} = V - gt$$

$$y = h + Vt - \tfrac{1}{2}gt^2.$$

To work out the maximum height achieved by the jumper, all we need to do is find out when $(dy/dt) = 0$ and see what value y has at this time.

When $(dy/dt) = 0$ we have $t = (V/g)$, and so

$$y_{\max} = h + \frac{V^2}{g} - \frac{1}{2}\frac{V^2}{g} = h + \frac{V^2}{2g}. \qquad (1.2.4)$$

To be able to use this result, we need to be able to give a value to h (the height of the centre of mass at take-off). The value of h is clearly going to be dependent on the position of the athlete's body at the moment of take-off (Fig. 1.12). To calculate the centre of mass we need a realistic,

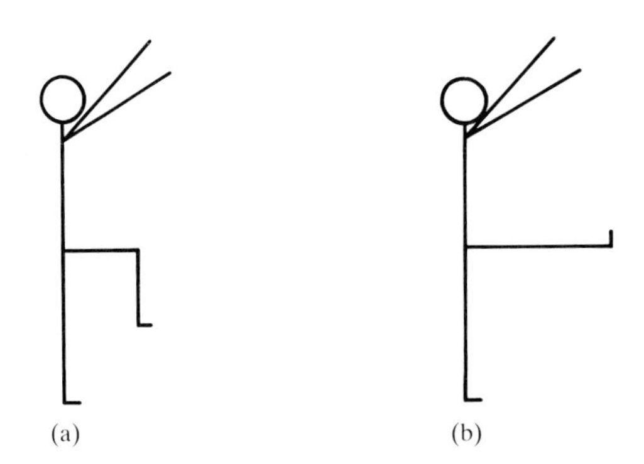

Fig. 1.12. The two standard positions at take-off.

yet simple, model of the athlete. Our model comprises seven hinged rods plus a disc for the head. Figure 1.13 shows the model with the lengths and masses of the various parts given as fractions of the total height (l) and total mass (m).

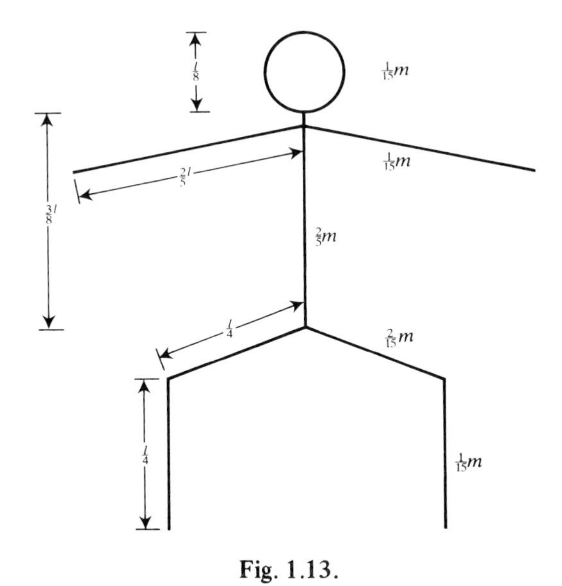

Fig. 1.13.

We now want to consider each of the positions shown in Fig. 1.12 in turn, to evaluate h. Figure 1.14 shows the various distances we will need.

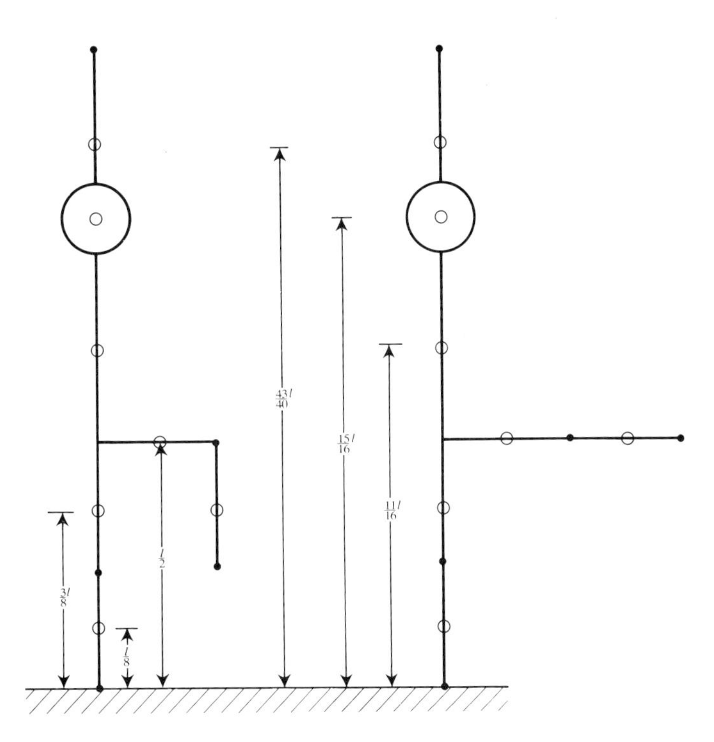

Fig. 1.14. The two standard positions at take-off.

In the position in Fig. 1.12(a), using the formula $\Sigma m_i x_i = M\bar{x}$ gives

$$mh = \underset{\text{Shin}}{\frac{m}{15}\cdot\frac{l}{8}} + \underset{\text{Thigh}}{\frac{2m}{15}\cdot\frac{3l}{8}} + \underset{\text{Thigh}}{\frac{2m}{15}\cdot\frac{l}{2}} + \underset{\text{Shin}}{\frac{m}{15}\cdot\frac{3l}{8}} + \underset{\text{Torso}}{\frac{2m}{5}\cdot\frac{11l}{16}} + \underset{\text{Head}}{\frac{m}{15}\cdot\frac{15l}{16}}$$

$$+ \underset{\text{Arms}}{\frac{2m}{15}\cdot\frac{43l}{40}}.$$

Thus $h = 0.631l$.

In the position in Fig. 1.12(b), a similar analysis gives

$$mh = \underset{\text{Shin}}{\frac{m}{15}\cdot\frac{l}{8}} + \underset{\text{Thigh}}{\frac{2m}{15}\cdot\frac{3l}{8}} + \underset{\text{Leg}}{\frac{m}{5}\cdot\frac{l}{2}} + \underset{\text{Torso}}{\frac{2m}{5}\cdot\frac{11l}{16}} + \underset{\text{Head}}{\frac{m}{15}\cdot\frac{15l}{16}} + \underset{\text{Arms}}{\frac{2m}{15}\cdot\frac{43l}{40}}$$

and $h = 0.639l$.

Since these values are so close we use $h = 0.635l$ from now on. We must next find how high the jumper needs to raise his centre of mass to clear a particular height of bar. Figure 1.15 shows a very simple model of the distance involved in calculating this for the case of the scissor jump.

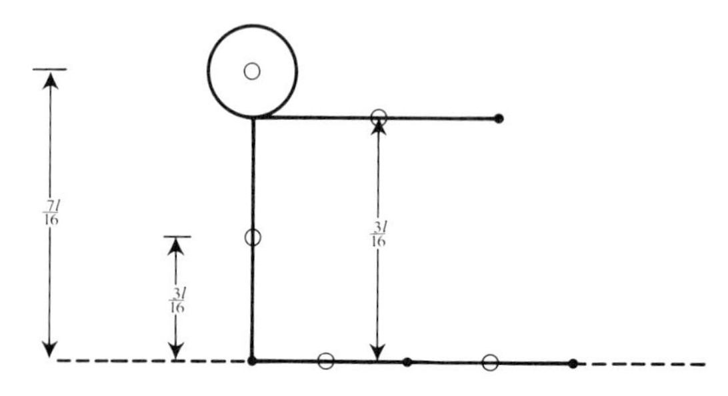

Fig. 1.15. The scissors jump.

The centre-of-mass formula gives

$$m\bar{y} = \underset{\text{Torso}}{\frac{2m}{5}\cdot\frac{3l}{16}} + \underset{\text{Arms}}{\frac{2m}{15}\cdot\frac{3l}{8}} + \underset{\text{Head}}{\frac{m}{15}\cdot\frac{7l}{16}},$$

so that the centre of mass is at a height $\bar{y} = 0.154l$ above the bar.

We can now use these figures to estimate the vertical impulse I at take-off. Equation (1.2.3) gives $I = mV$, and eqn (1.2.4) for $y_{\max}$ gives

$$V = m\{2g(y_{\max} -- h)\}^{\frac{1}{2}}\ ;$$

putting these together we have

$$I_{\mathrm{SC}} = m\{2g(y_{\max} - h)\}^{\frac{1}{2}}$$
$$= m\{2g(H + 0.154l - 0.635l)\}^{\frac{1}{2}}$$

where H is the height of the bar.

For a 1.94 m tall high jumper, clearing a bar at a height of 2.10 m using a scissor jump, we have

$$I_{\mathrm{SC}} = m\{2g(1.17)\}^{\frac{1}{2}}.$$

In the straddle jump, the athlete's body is horizontal and parallel to the bar as the bar is cleared and so, for simplicity, we assume that his centre of mass is at the height of the bar. We then have

$$I_{\mathrm{ST}} = m\{2g(y_{\max} - h)\}^{\frac{1}{2}}$$
$$= m\{2g(H - 0.635l)\}^{\frac{1}{2}}.$$

For our 1.94 m jumper with the bar at 2.10 m this gives

$$I_{ST} = m\{2g(0.87)\}^{\frac{1}{2}}.$$

Finally we examine the Fosbury flop. Figure 1.16 shows the position of the athlete as he clears the bar. Choosing the hips as the origin as shown in Fig. 1.16, the coordinates of the centre of mass are given as follows (lengths and masses can be taken from Fig. 1.13)

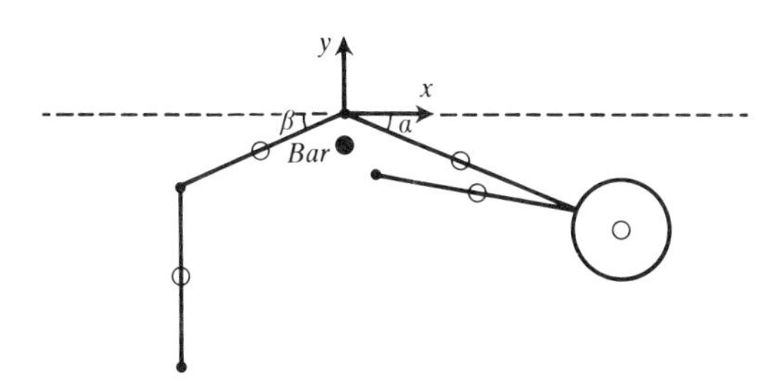

Fig. 1.16. The Fosbury flop.

$$m\bar{x} = -\frac{14m}{15}\cdot\frac{l}{8}\cos\beta - \frac{2m}{15}\cdot\frac{l}{4}\cos\beta + \frac{2m}{5}\cdot\frac{3l}{16}\cos\alpha + \frac{2m}{15}\cdot\frac{7l}{40}\cos\alpha + \frac{m}{15}\cdot\frac{7l}{16}\cos\alpha.$$

Thighs Shins Torso Arms Head

Thus $\bar{x} = (0.1275\cos\alpha - 0.0667\cos\beta)l$ and

$$m\bar{y} = \frac{-4m}{15}\cdot\frac{l}{8}\sin\beta - \frac{2m}{15}\left(\frac{l}{4}\sin\beta + \frac{l}{8}\right)\frac{2m}{5}\cdot\frac{3l}{16}\sin\alpha - \frac{2m}{15}\cdot\frac{7l}{40}\sin\alpha - \frac{m}{15}\cdot\frac{7l}{16}\sin\alpha.$$

Thighs Shins Torso Arms Head

Thus $\bar{y} = -(0.1275\sin\alpha + 0.0667\sin\beta + 0.0167)l$.

For our 1.94 m tall high jumper and α and β both equal to 20° we have

$$(\bar{x}, \bar{y}) = (0.11, -0.16)$$

and so the centre of mass is (roughly) 0.16 m *below* the bar! We can now mirror the calculation above to get the impulse in this case as

$$I_F = m\{2g(y_{max} - h)\}^{\frac{1}{2}}$$

$$= m\{2g(H - 0.16 - 0.635l)\}^{\frac{1}{2}}$$

$$= m(2g)^{\frac{1}{2}}(0.84).$$

The ratio of the impulses for the different jumping techniques is then given by

$$I_{SC} : I_{ST} : I_F = 1.17 : 0.87 : 0.84$$

$$= 1 : 0.74 : 0.72.$$

The results show that the Fosbury flop technique is the most efficient since the raising of the centre of mass is the most important factor of high jumping.

• Problems

1. Compare the efficiency of the scissor and straddle jumps with the bar at a height of 1.40 m.

2. Compare the impulses required by two Fosbury floppers of heights 1.90 m and 1.94 m, respectively, when trying to clear a height of 2.15 m.

• Further reading

Open University MST204 Summer School Video Cassette, 1983.
Townend, M. S. (1984). *Mathematics in sport*. Ellis Horwood, Chichester.

1.3. Down-hill skiing

By what means can a skier increase his overall speed for a down-hill ski course? The aim of increasing overall speed, or, equivalently, decreasing the course time, can be accomplished by altering certain factors which are within the skier's control. For example, skiers constantly wax their skis with specially prepared waxes so as to decrease the friction between the skis and the snow as much as possible.

Our aim is to formulate a mathematical expression describing the motion of a skier on a down-hill ski run, and from this we hope to be able to detect factors which can be altered to enable us to increase the skier's speed.

The conditions which affect the skier's descent, and which we will take into account, are

1. The component of gravity acting down the slope;
2. Friction between the snow and the skis;
3. Air resistance.

Many other relevant features will not be incorporated in our formulation of the skier's motion for the sake of simplicity. The reader should consider the possibilities, and decide on the significance of these omissions. For example, the skier is able to aid the effects of gravity by the use of his ski sticks, but, apart from increasing his stability, is the effect on his speed by pushing backwards very great?

We shall produce a mathematical description of the descent in three stages.

(a) Simplify the conditions (1), (2), and (3) above, and find mathematical expressions for them.

(b) Use Newton's second law of motion, incorporating conditions (1), (2), and (3), to construct an equation of motion for the skier's descent. From this equation we shall find an expression for the 'terminal velocity' of the skier (i.e. the maximum velocity reached during the descent which, once reached, is maintained).

(c) From the equation of motion we shall produce an expression relating the skier's velocity to the distance he has travelled at any point during the descent. This will involve the ability to solve a first-order differential equation by the 'integrating factor' method.

• Solution

(a) SIMPLIFYING CONDITIONS 1, 2, AND 3

Gravity

Gravity acts on the skier's total mass m (including skis) with a vertical force mg. The component of this force which pulls the skier down the slope is $mg \sin \alpha$, if we *assume* that the ski-run is a constant slope of angle α. In fact α will vary during the length of the run, but we can approximate the course to a plane, inclined at α to the horizontal, where α can be calculated from the difference in altitude at the start and finish, and the length of the course.

Friction

The friction F opposes the motion. We know that $F = \mu N$, where μ is the coefficient of dynamic friction for skis on snow and N is the normal reaction to the snow surface. Resolving forces normal to the plane in the diagram in Fig. 1.17 we have $N = mg \cos \alpha$ and thus $F = \mu mg \cos \alpha$.

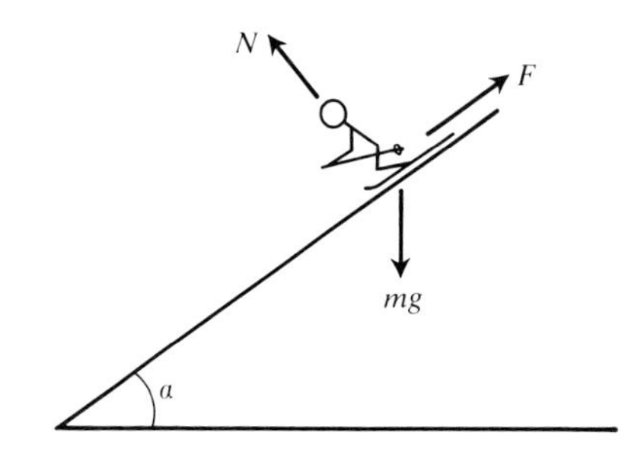

Fig. 1.17. Forces on skier neglecting air resistance.

Air resistance

Air resistance for drag, like friction, opposes the motion. Three factors affect the drag D.

1. The velocity of the skier v;
2. The area of the skier facing the resisting air A;
3. The density of the air ρ.

Experimentally D is found to obey

$$D \propto v^2; \quad D \propto A; \quad D \propto \rho.$$

Combining these gives $D \propto v^2 A \rho$.

By convention we shall introduce a constant $\frac{1}{2}C_D$, where C_D is called the drag coefficient. Thus,

$$D = \tfrac{1}{2}C_D v^2 A \rho.$$

(b) FINDING THE TERMINAL VELOCITY

Figure 1.18 shows friction, gravity, and drag acting on the skier. If a is the acceleration down the slope, Newton's second law of motion says that ma must equal the resultant force down the plane, i.e. $ma = P - D - F$.

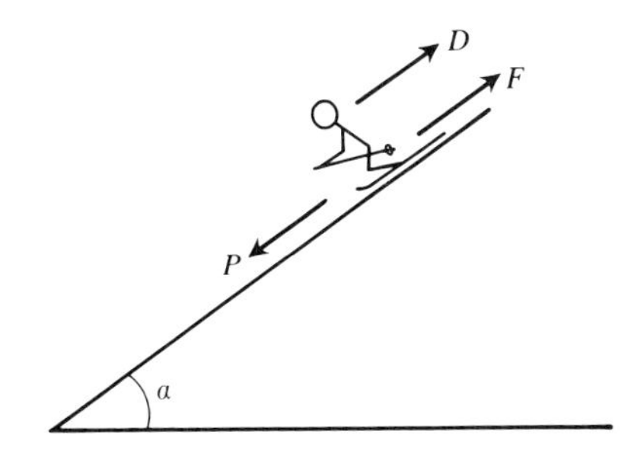

Fig. 1.18. Forces on skier including air resistance.

Substituting the expression for P, D, and F we have the required equation of motion

$$ma = mg \sin \alpha - \tfrac{1}{2} C_{\mathrm{D}} v^2 A \rho - \mu mg \cos \alpha. \qquad (1.3.1)$$

The terminal velocity of an object is reached if its acceleration ceases, i.e. $a = 0$. Letting $a = 0$ in eqn (1.3.1), we have

$$0 = mg \sin \alpha - \tfrac{1}{2} C_{\mathrm{D}}\, v_T A \rho - \mu mg \cos \alpha$$

where v_{T} is the terminal velocity.

Hence

$$v_{\mathrm{T}} = \left(\frac{2mg}{C_{\mathrm{D}}\, A\rho} (\sin \alpha - \mu \cos \alpha) \right)^{\frac{1}{2}}. \qquad (1.3.2)$$

Before we attempt stage (c) let us substitute some values into eqn (1.3.2) to find a rough value for the terminal velocity of a skier.

Take $m = 80$ kg; $g = 9.81$ m s^{-2}; $\rho = 1.25$ kg m^{-3}; $C_{\mathrm{D}} = 0.7$ (this is an experimental value for the human body); $\alpha = 12^{\circ}$; and $\mu = 0.05$ (for well-waxed skis). We still need a value for A.

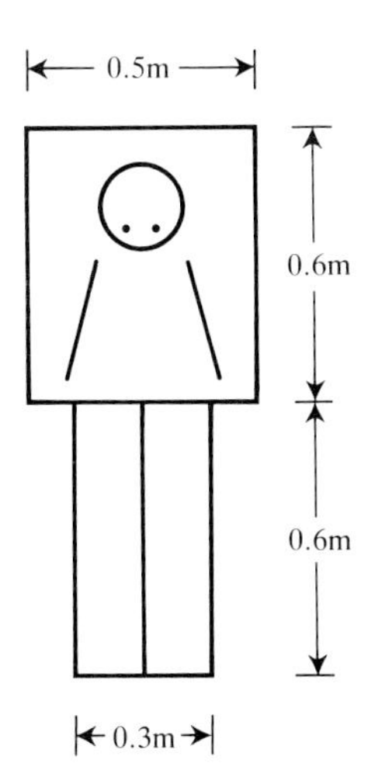

Fig. 1.19

Figure 1.19 gives an abstract view of the front of skier in the crouched position. The area of this figure will give us an approximate value of A. Thus $A = (0.5 \times 0.6) + (0.3 \times 0.6) = 0.48\ \mathrm{m}^2$. Substituting this value and the other constants in eqn (1.3.2) gives $v_\mathrm{T} = 24.38\ \mathrm{m\,s^{-1}}$, or about 54.5 m.p.h.

(c) THE RELATIONSHIP BETWEEN THE SKIER'S VELOCITY AND DISTANCE TRAVELLED

Let the distance travelled down the slope from the start at any time t (measured in seconds) be s (measured in metres), then we can write

$$a = \frac{\mathrm{d}v}{\mathrm{d}t} = \frac{\mathrm{d}v}{\mathrm{d}s}\cdot\frac{\mathrm{d}s}{\mathrm{d}t} = \frac{\mathrm{d}v}{\mathrm{d}s}\cdot v.$$

Note: the above is true since

$$\frac{\mathrm{d}v}{\mathrm{d}t} = \frac{\mathrm{d}v}{\mathrm{d}s}\cdot\frac{\mathrm{d}s}{\mathrm{d}t},$$

using the 'function of a function' rule and $\mathrm{d}s/\mathrm{d}t = v$.

Substituting $a = (\mathrm{d}v/\mathrm{d}s)\cdot v$ in eqn (1.3.1) gives

$$Mv\frac{\mathrm{d}v}{\mathrm{d}s} = mg\sin\alpha - \tfrac{1}{2}C_\mathrm{D}v^2A\rho - \mu mg\cos\alpha$$

$$mv\frac{\mathrm{d}v}{\mathrm{d}s} + \tfrac{1}{2}C_\mathrm{D}v^2A\rho = mg(\sin\alpha - \mu\cos\alpha)$$

$$v\frac{\mathrm{d}v}{\mathrm{d}s} + \frac{C_\mathrm{D}v^2A\rho}{2m} = g(\sin\alpha - \mu\cos\alpha).$$

Now note that

$$\frac{\mathrm{d}v}{\mathrm{d}s}(\tfrac{1}{2}v^2) = v\frac{\mathrm{d}v}{\mathrm{d}s}$$

(again by the 'function of a function' rule), and hence

$$\frac{\mathrm{d}}{\mathrm{d}s}(\tfrac{1}{2}v^2) + \frac{C_\mathrm{D}A\rho}{m}(\tfrac{1}{2}v^2) = g(\sin\alpha - \mu\cos\alpha). \tag{1.3.3}$$

This equation is a first-order linear differential equation which can be solved by the integrating factor method treating the variable as $\tfrac{1}{2}v^2$.

The integrating factor is $I = \mathrm{e}^{\int C_\mathrm{D}A\rho\,\mathrm{d}s/m} = \mathrm{e}^{C_\mathrm{D}A\rho s/m}$.

$$v = +\left[\frac{2mg(\sin\alpha - \mu\cos\alpha)}{C_\mathrm{D}A\rho}(1 - \mathrm{e}^{-C_\mathrm{D}A\rho s/m})\right]^{1/2} \tag{1.3.4}$$

or since

$$v_\mathrm{T} = \left[\frac{2mg(\sin\alpha - \mu\cos\alpha)}{C_\mathrm{D}A\rho}\right]^{1/2}$$

$$v = +(1 - \mathrm{e}^{-C_\mathrm{D}A\rho s/m})^{1/2}v_\mathrm{T} \tag{1.3.5}$$

The equivalent results (1.3.4) and (1.3.5) give us a value of v in terms of known constants and s. Let us, as before for v_T, substitute values into one of these equations to find a sample velocity at a given distance.

We shall choose all the values to be the same as before where v_T was 24.38 m s^{-1} and choose $s = 200$ m. From eqn (1.3.5) these give $v = 15.85$ m s^{-1} or about 35.5 m.p.h.

ANALYSIS OF RESULTS

Our aim was to find ways of allowing a skier to ski as fast as possible. We now have a result which shows us his speed at any distance down the slope from the start, i.e.

$$v = +[1 - e^{-C_D A\rho s/m}]^{1/2} v_T$$

or

$$v = +[1 - e^{-C_D A\rho s/m}]^{1/2} \left[\frac{2mg(\sin\alpha - \mu\cos\alpha)}{C_D A\rho}\right]^{1/2}. \quad (1.3.6)$$

For a fixed slope α, the only constants which can be altered are C_D, A, m, and μ.

(i) A change in μ

Since μ occurs only in the right-hand part of the expression for v (eqn (1.3.1)) and its reduction will cause this to increase, v in turn will also increase. This is to be expected, since lowering friction should increase speed.

Let us suppose the skis are specially prepared such that it is possible to reduce μ to 0.04. What would the speed at the 200 m mark now be (it was 15.85 m s^{-1}), assuming all other constants remain as before?

Using eqn (1.3.6) we have $v = 16.33$ m s^{-1}. This is a reasonable increase, but it is very difficult to reduce μ much further than 0.05 in practice.

(ii) A change in C_D, m, or A

If we write $k = C_D A/m$, then eqn (1.3.6) becomes

$$v = +[1 - e^{-k\rho s}]^{1/2} \left[\frac{2g(\sin\alpha - \mu\cos\alpha)}{k}\right]^{1/2}. \quad (1.3.7)$$

Suppose k as a whole is decreased what will happen to v?

Before, at $s = 200$ m, when v was 15.85 m s^{-1}, k was $(0.7 \times 0.48)/80 = 4.2 \times 10^{-3}$ (μ is again 0.05). Let us try reducing k by 10 per cent, i.e. to 3.78×10^{-3}. v is now 22.46 m s^{-1}. This is a significant increase. k can be decreased either by decreasing C_D or A, or by increasing m. Thus the skier has the following options.

1. To reduce C_D. This can be done by streamlining himself: the only practicable way being to make his ski-suit more slippery and shiny.
2. To reduce A. This is done by crouching as low as possible and by keeping the arms in front close to the body when not being used.
3. To increase m. This can be achieved by either 'weighting' the skier or more probably the skis.

SUMMARY

For most of our discussion calculations have been made at a specific distance (i.e. $s = 200$ m). It really would not have mattered what s had been, and our conclusion that reduction of μ, A, or C_D and increase of m increases v thus means that v is increased over the whole course and the overall time for the ski-run is reduced.

• Problems

1. Using the established formulae calculate (a) the terminal velocity and (b) the velocity at a point 400 m from the start for a skier descending a run of average inclination $\alpha = 15°$, and where $\rho = 1.25\ \text{kg m}^{-3}$, $C_D = 0.7$, $A = 0.5\ \text{m}^2$, $m = 75$ kg, and $\mu = 0.05$.

2. Explain why the terminal velocity is never reached.

3. The total time to cover a course of length s can be found by integrating eqn (1.3.7) with respect to time. The result is again a function of s. Find this expression for T, the total time taken. Hint: If v is written as ds/dt, eqn (1.3.7) is of the variable separable type. The substitution $z^2 = 1 - e^{-C_D A \rho s/m}$ and then the use of partial fractions will help in finding the value of the required integral.

4. Find the time required to complete a 2000 m long ski-run using the data supplied in Problem 1.

5. A ski equipment manufacturer wishes to improve his products with the aim of advertising a claim that one of his products will allow a skier to ski 1 per cent faster than with 'normal' equipment. Looking at eqn (1.3.7), it can be seen that the only factors a manufacturer can have control over are C_D and μ (the others are either fixed or governed by the skier himself). He can decrease the former by developing better ski clothing, and the latter by producing a smoother surface to skis or better waxes. Roughly what percentage decrease is required in (i) C_D and (ii) μ in order to decrease v at any point by 1 per cent?

6. A skier descending a run at an average inclination of 15° collides with a tree after 1000 m! If his mass (clothed, with skis) is 80 kg, and $A = 0.5\ \text{m}^2$, $\rho = 1.25\ \text{kg m}^{-3}$, $\mu = 0.06$, and $C_D = 0.7$, find the impact of the tree on the skier (i.e. impulse). From what height, dropping vertically on to solid ground, would a person have to drop to suffer the same impact?

• Further reading

Townend, S. M. (1984). *Mathematics in sport*. Ellis Horwood, Chichester.

1.4. The shot-putter's problem

Projectiles

A shot-putter concentrates on a smooth build-up of the speed of his body across the circle and extending his arm, thus accelerating the shot in order to give maximum momentum of the shot at the point of release. He starts in a crouched position with the shot resting under his chin and finishes in an upright position with arm fully extended. The angle that the extended arm makes with the horizontal will affect the angle of projection of the shot; also the putter 'explodes' into action across the circle and the speed of projection of the shot depends on this initial burst. The putter thus controls the speed and angle of the shot.

Clearly the shot-putter's problem is to launch the shot in such a way that the range is as long as possible. Figure 1.20 illustrates the motion of the shot. A shot-putter usually puts the

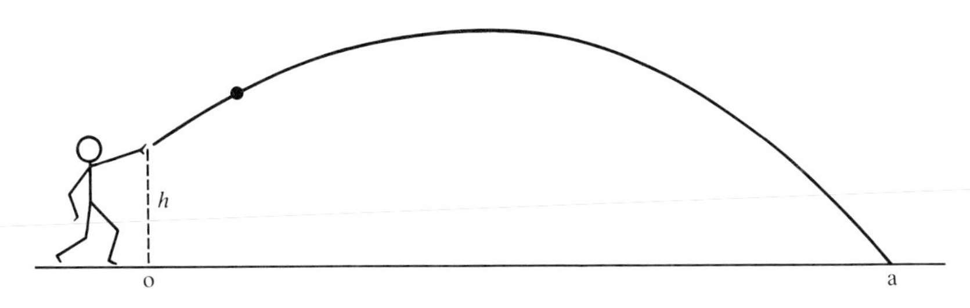

Fig. 1.20

shot with the same speed every time and this depends on his physical strength and style. If a world champion launches the shot from a height of 2 metres and achieves a range (A0) of 22.15 metres, what is the angle and speed of release of the shot?

• Solution

Suppose that the angle of release is α and the speed of launch is u (Fig. 1.21). Then, using standard projectile theory and neglecting air resistance, the equation of the path of the shot is given by

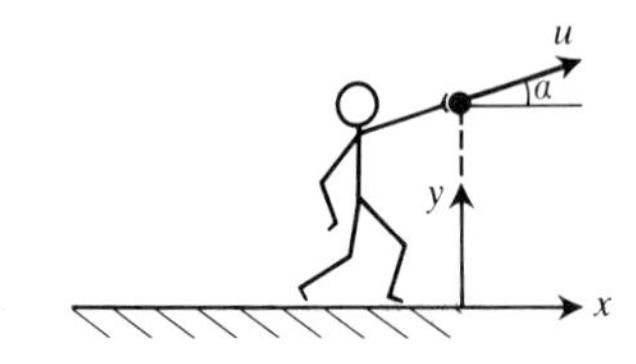

Fig. 1.21

$$y = x \tan \alpha - \frac{gx^2}{2u^2 \cos^2 \alpha} + h.$$

If we let R be the range of the shot then R satisfies the equation,

$$0 = R \tan \alpha - \frac{gR^2}{2u^2 \cos^2 \alpha} + h.$$

Writing $1/\cos^2 \alpha$ as $\sec^2 \alpha$ and rearranging, this becomes

$$R \tan \alpha - \frac{g}{2u^2} \sec^2 \alpha R^2 + h = 0. \qquad (1.4.1)$$

The maximum value of R occurs when $(\mathrm{d}R/\mathrm{d}\alpha) = 0$. Differentiating with respect to α, we have

$$\frac{\mathrm{d}R}{\mathrm{d}\alpha} \tan \alpha + R \sec^2 \alpha - \frac{g}{2u^2}(2 \sec^2 \alpha \tan \alpha) R^2 - \frac{g}{2u^2} \sec^2 \alpha \frac{2R\mathrm{d}R}{\mathrm{d}\alpha} = 0,$$

but $(\mathrm{d}R/\mathrm{d}\alpha) = 0$ at the maximum so that this equation reduces to

$$R \sec^2 \alpha \left(1 - \frac{Rg}{u^2} \tan \alpha\right) = 0,$$

i.e.

$$\tan \alpha = u^2/Rg. \qquad (1.4.2)$$

Substituting for $\tan \alpha$ and $\sec^2 \alpha (= 1 + \tan^2 \alpha)$ in eqn (1.4.1) and solving for R, we get

$$R = \frac{u^2}{g}\left(1 + \frac{2gh}{u^2}\right)^{\frac{1}{2}}. \qquad (1.4.3)$$

Note that, if $h = 0$, then $R = (u^2/g)$ and $\tan \alpha = 1$, which are the results of standard projective theory.

Equation (1.4.3) can be rearranged to give

$$u^2 = g\{(h^2 + R^2)^{\frac{1}{2}} - h\}. \qquad (1.4.4)$$

Using this value for u^2 in eqn (1.4.2) gives

$$\tan \alpha = \frac{(h^2 + R^2)^{\frac{1}{2}} - h}{R} \qquad (1.4.5)$$

Putting $h = 2$ m, $g = 9.8$ m s^{-2}, and $R = 22.15$ m in eqns (1.4.4) and (1.4.5) gives

$$u \simeq 14.08 \text{ m s}^{-1} \quad \text{and} \quad \alpha \simeq 42.2°.$$

A much simpler (although not very realistic) problem can be posed in which the shot is assumed to be launched from zero height. The solution then involves the maximum range $L = u^2/g$ and a launch angle of 45°. This simplified problem can be used to investigate the practical question—whether it is better to concentrate on achieving the optimum angle of launch or to go for an increased launch speed. This is discussed in the further problems.

• Problems

1. If we assume that the shot is launched at ground level, then the range R is related to the launch speed u and angle α by $R = u^2 \sin 2\alpha/g$. The optimum angle is now 45° and, for a maximum range 22.15 metres, we have $u = 14.73$ m/s. Investigate the changes in the range R by

    (i) varying α but keeping u fixed;

    (ii) varying u but keeping α fixed,

and hence advise the shot-putter on whether to concentrate on achieving the optimum angle of projection or trying to increase the speed of projection.

2. Repeat the problem carrying out your own experiments with a shot. You will need to know (i) how far you can put (not throw!) and show (ii) at what height above the ground you launch the shot and (iii) the angle of launch.

From this data you can calculate your speed of launch for the shot. Hence calculate your maximum range and optimum launch angle for the shot. Try it out on the sports field. Will you break any records?

3. The models formulated for this problem have ignored the effects of air resistance. For a spherical object of diameter D metres and speed u m s^{-1}, experiments in a wind tunnel show that the magnitude of the force due to air resistance is approximately equal to $0.2D^2u^2$ newtons. By comparing this with the magnitude of the other forces acting on the shot, decide whether our decision to ignore air resistance was justified.

4. In a competition the qualifying distance is 20 metres. Using the model given by the equation

$$y = x \tan \alpha - \frac{gx^2}{2u^2 \cos^2 \alpha} + h$$

where $h = 2$ m, $g = 9.8$ m s^{-2}, and the initial speed is 14 m s^{-1}, find the range of values of α which will result in a qualifying throw.

• Further reading

Berry, J. S. and Young, B. (1982). Modelling a shot putt. In *Solving real problems with mathematics*, Vol. I, The Spode Group. Cranfield Press, p. 98.

1.5. The long golf drive
Impulse; momentum; collisions

As every budding Sevvy Ballesteros knows, the tee shot on each hole in a round of golf plays a crucial part in obtaining a low score for the round. Not only does it give a satisfying feeling to hit the ball consistently far down the fairway, but a drive of over 200 metres usually means that half the hole has been 'conquered' in one shot. The top professional golfers expect to drive distances over 300 metres; however, the average high handicap golfer (i.e. a handicap of around 18–24) will be happy with drives of 180–200 metres.

So what are some of the features which affect the length of a drive? When the ball has left the tee, factors such as wind, the lie of the land, and the hardness of the fairway become important. One overriding factor in obtaining a good drive must be the achievement of high initial velocity of the ball off the tee. This will depend on the size and mass of the clubhead as well as its speed and point of contact on impact with the ball. The size, mass, and elasticity of the golf ball are also relevant. The dimensions of the golf ball are fixed by the golf legislators; the mass is 46 grams and the diameter is 1.68 inches (4.26 cm).

How does the speed of projection depend on the clubhead speed and mass?

• Solution

Consider a very simplified model in which the clubhead has mass M and speed U just before a horizontal impact and speed v after impact; and that the golf ball has mass m and speed w after impact (Fig. 1.22). Then the law of conservation of linear momentum gives

$$MU = Mv + mw.$$

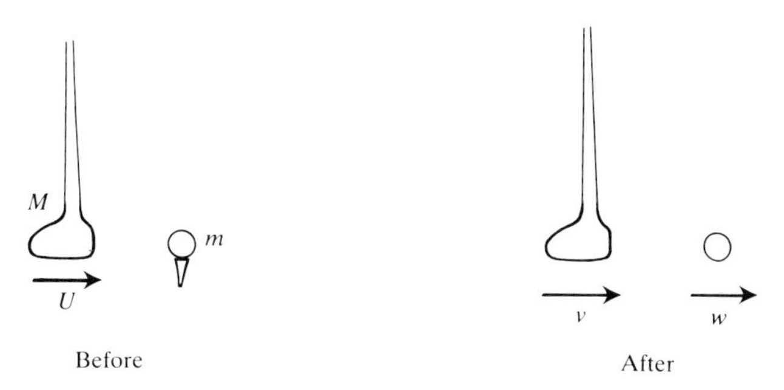

Fig. 1.22

If we assume Newton's experimental law for impacts, then

$$e = \frac{w - v}{U}$$

where e is the coefficient of restitution between the ball and the clubhead. Eliminating v gives

$$w = \frac{M(1 + e)U}{M + m}. \qquad (1.5.1)$$

Now m is fixed, so eqn (1.5.1) shows how the initial speed of the ball w depends on U, M, and e.

From eqn (1.5.1) it is clear that w depends linearly on U so that a 5 per cent increase in U will give a 5 per cent increase in w. This fact is probably intuitively obvious. However, the value of w can also be increased by changing M and e.

Figure 1.23 shows a graph of $M/(M + m)$ plotted against M using a value of $m = 46$ g. We see that, if the clubhead mass is below about 150 g, a fair amount of improvement can be made by increasing its mass. However, when M reaches around 400 g, an increase in mass causes

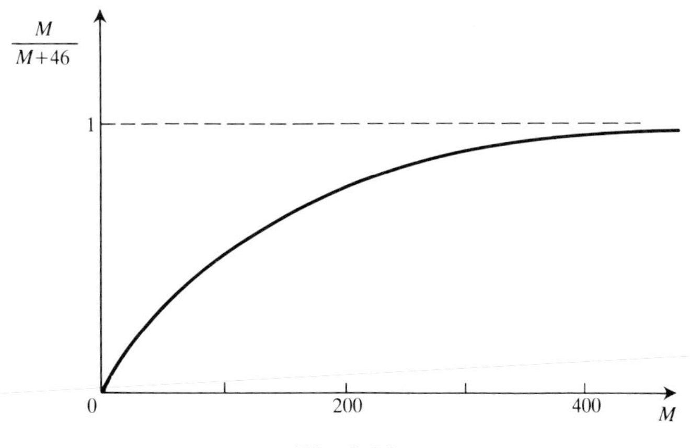

Fig. 1.23

a negligible increase in $M/(M + 46)$ and hence speed w. Another important point to the golfer is that a very heavy club may prove difficult to lift and swing at high speed. Thus the benefits of an increase in mass may be lost due to a decrease in initial impact speed U.

The elasticity of the ball is another factor to consider. The model predicts that w varies linearly with e. A typical value for e for a golf ball and clubhead collision is 0.7; this value is for a good quality golf ball—the lower the quality, the smaller is the value of e in general. Although this value is fairly consistent for a large range of velocities, it does change with temperature and over the range 0–27 °C, e may vary from 0.64 to 0.75. It is thus important to keep the ball warm!

• Problems

1. If a clubhead speed of 160 km/h imparts an initial speed of 216 km/h to a golf ball, find the approximate mass of the clubhead M taking the value of e to be 0.7 and $m = 46$ g. Using this value for M draw a graph showing how the initial speed of the golf ball varies with e for values of e between 0.65 and 0.75.

2. A golfer's swing gives a clubhead speed of 38 m/s when the clubhead has a mass of 100 g but this drops to 34 m/s when the clubhead has a mass of 300 g. It is assumed that there is a linear relation connecting the clubhead speed U with its mass M of the form

$$U = a\left(1 - \frac{M}{b}\right)$$

where a and b are constants.

Find a and b.

Thus, use the model developed in this chapter to find the mass of the clubhead which would optimize this golfer's drive, given that the ball's mass is 50 g.

Sketch a graph of w, the initial speed of the golf ball, against M for the range $100 \leqslant M \leqslant 300$.

3. A good approximation to the maximum range R which a golf ball can carry, for projection speeds w between 30 and 70 m s^{-1}, is given by $R = w^2/2g$, i.e. half the maximum range from projectile theory.

4. A golfer who played throughout the year put down his apparent loss of form off the tee in cold weather to the restriction of wearing extra clothes to keep warm. Assume, in fact, that he achieves approximately the same clubhead speed of 44 m s^{-1} with a clubhead mass of 250 g throughout the year. If he uses a ball of mass 50 g, compare his drives on a freezing winter day and on a hot summer day.

1.6. The ice skater's problem

Moment of inertia; angular momentum; rotational energy

An ice skater going into a spin starts spinning with the arms horizontal and increases the rate of spin by letting the arms fall to a vertical position. The problem is to have a rate of spin at the start which will give a satisfactory rate of spin when the arms are vertically down.

• Solution

We first need a physical model of the human body. A simple model is to consider it to be made up from three cylinders—one for each of the two arms and one to represent the head, torso, and legs (Fig. 1.24).

In order to discover the relationship between the rate of spin at the start and that after the arms have been dropped it is necessary to consider the kinetic energies of rotation of the body in the initial and final positions. These will be the same by the conservation of energy principle.

The kinetic energy of a rotating body is given by

$$\tfrac{1}{2}Iw^2$$

The ice skater's problem

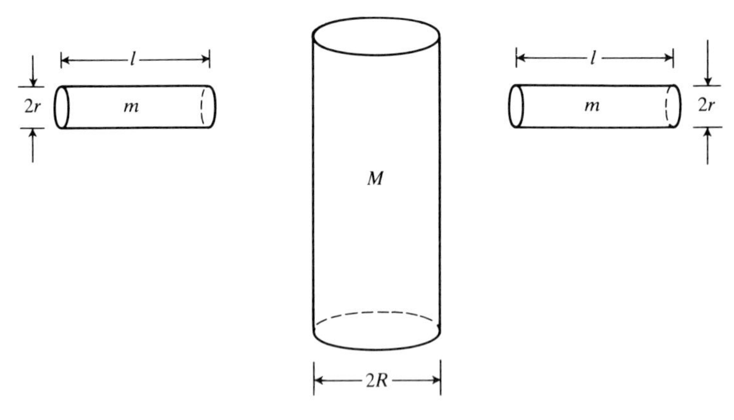

Fig. 1.24

where I = moment of inertia and w = angular velocity. Figure 1.25 shows that initially the centre of mass of each arm is at a distance $\frac{1}{2}l + R$ from the body centre line. A table of standard results gives the moment of inertia for each arm when horizontal as

$$I_{xx'} = m\left(\frac{r^2}{4} + \frac{l^2}{12}\right)$$

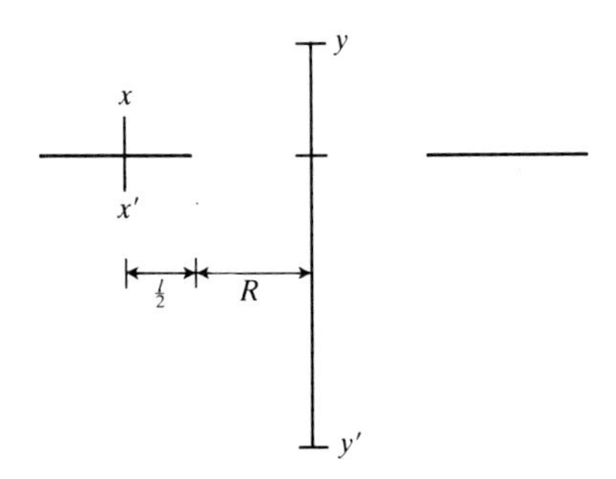

Fig 1.25

and, using the parallel-axes theorem,

$$I_{yy'} = m\left[\frac{r^2}{4} + \frac{l^2}{12} + \left(\frac{l}{2} + R\right)^2\right]$$

for each arm. For the trunk, tables of moment of inertia quote

$$I_{yy'} = \tfrac{1}{2}MR^2,$$

so for the whole body we have

$$I_0 = 2m\left[\frac{r^2}{4} + \frac{l^2}{12} + \left(\frac{l}{2} + R\right)^2\right] + \tfrac{1}{2}MR^2.$$

When the arms are vertically downwards a similar analysis gives

$$I_1 = 2m\{\tfrac{1}{2}r^2 + (r + R)^2\} + \tfrac{1}{2}MR^2.$$

By the principle of conservation of energy, if w_0 is the initial angular velocity and w_1 is the final angular velocity, then

$$\tfrac{1}{2}I_0 w_0^2 = \tfrac{1}{2}I_1 w_1^2.$$

So

$$\frac{1}{2}\left[2m\left(\frac{r^2}{4}+\frac{l^2}{12}+\left(\frac{l}{2}+R\right)^2\right)+\tfrac{1}{2}MR^2\right]w_0^2$$

$$=\frac{1}{2}\left[2m\;\frac{r^2}{2}+(r+R)^2)+\tfrac{1}{2}MR^2\right]w_1^2.$$

Suppose $m = 3$ kg; $r = 0.05$ m; $l = 0.75$ m; $M = 60$ kg; $R = 0.15$ m; $h = 1.8$ m. We then have

$$1.31w_0^2 = 0.46w_1^2.$$

Thus, in ratio notation,

$$w_1 : w_0 = 1.68.$$

Hence if the skater's initial rotation is 2.5 revolutions per second, the final rotation is 4.2 revolutions per second.

Comment

It would be useful to study a video recording of an ice skater and compare the theoretical results with actual performances in order to see if a better model can be achieved. It may be a good idea to study this problem when an ice skating competition is on the television.

Experimental results may be obtained by mounting a car wheel on a vertical axis. A pupil can then stand on the wheel with arms horizontal and be set spinning. If the pupil lowers his arms, the two speeds of rotation may be compared.

• Problems

1. By measuring other pupils, compare the results for pupils of different physiques. What is the best physique for this particular activity?

2. Repeat the problem with the skater raising the arms vertically above the shoulders.
A better physical model is to use five cylinders—two for the arm, two for the legs, and one for the torso. Try this, to see whether it changes the final $w_1 : w_0$ ratio substantially.

• Further reading

Townend, M. S. (1984). *Mathematics in sport*. Ellis Horwood, Chichester.

1.7. Miracle at Mexico City?

The 1968 Olympic Games, held at Mexico City, saw the world record for the men's long jump raised from 8.35 to 8.90 m—a staggering increase of almost 10 per cent. How staggering is illustrated by the fact that it is still the world record today! The list of world record-breaking jumps since 1901 is given in Table 1.1.

Many commentators suggested that the winning jump by Bob Beamon was not really as miraculous as it appeared to be, arguing that the reduced air density at the high altitude of Mexico City (2256 m above sea level) would considerably enhance the jump. An obvious counter-argument to this was that the other jumpers' performances, if anything, were worse than usual. Looking at the first six places we have the final results.

Bob Beamon	8.90 m
Klaus Beer	8.19 m
Ralph Boston	8.16 m
Igar Ter-Ovanesyan	8.12 m
Tonu Lepik	8.09 m
Allen Crawley	8.02 m

All the jumps but that of Bob Beamon were considerably less than the current world record of 8.35 m. The other competitors may of course have been discouraged by Beamon's first jump, so determine just what effect air resistance had on Beamon's jump.

Table 1.1. Long jump—men

Length of jump	Competitor	Country	Date	Home of competitor
7.61 m/24'11$\frac{3}{4}$"	Peter O'Connor	GBR	5.8.1901	Dublin
7.69 m/25'2"	Edwin Gourdin	USA	23.7.1921	Cambridge, Mass.
7.76 m/25'5$\frac{1}{4}$"	Robert LeGendre	USA	7.7.1924	Paris
7.89 m/25'10$\frac{3}{4}$"	William de Hart Hubbard	USA	13.6.1925	Chicago
7.90 m/25'11"	Edward Hamm	USA	7.7.1928	Cambridge, Mass.
7.93 m/26'0$\frac{1}{4}$"	Sylvio Cator	HAI	9.9.1928	Paris
7.98 m/26'2$\frac{1}{4}$"	Chuhei Nambu	JAP	27.10.1931	Tokyo
8.13 m/26'8$\frac{1}{4}$"	Jesse Owens	USA	25.5.1935	Ann Arbor, Mich.
8.21 m/26'11$\frac{1}{4}$"	Ralph Boston	USA	12.8.1960	Walnut, Calif.
8.24 m/27'0$\frac{1}{2}$"	Ralph Boston	USA	27.5.1961	Modesto, Calif.
8.28 m/27'2"	Ralph Boston	USA	16.7.1961	Moscow
8.31 m/27'3$\frac{1}{4}$"	Igor Ter-Ovanesyan	URS	10.6.1962	Yerevan
8.31 m/27'3$\frac{1}{4}$"	Ralph Boston	USA	15.8.1964	Kingston, Jamaica
8.34 m/27'4"	Ralph Boston	USA	12.9.1964	Los Angeles
8.35 m/27'5"	Ralph Boston	USA	29.5.1965	Modesto, Calif.
8.35 m/27'4$\frac{3}{4}$"	Igor Ter-Ovanesyan	URS	19.10.1967	Mexico City
8.90 m/29'2$\frac{1}{2}$"	Bob Beamon	USA	18.10.1968	Mexico City

• Solution

If we model the situation by assuming that the jumper is a projectile moving with velocity v, so that the speed v is given by $v = |\mathbf{v}|$, then Newton's second law of motion gives

$$\frac{m\,\mathrm{d}\mathbf{v}}{\mathrm{d}t} = m\mathbf{g} + \mathbf{D} \qquad (1.7.1)$$

where m is the jumper's mass, $\mathbf{g} = -g\mathbf{j}$, and $\mathbf{D}$ is the drag force on the body (Fig. 1.26(a)). It is usual to assume that the drag force is proportional jointly to the square of the speed v^2 and the density of air ρ. If $\mathbf{v}$ is the velocity vector the drag force is in the negative $\mathbf{v}$ direction, of magnitude $k\rho v^2$ and so can be written as

$$\mathbf{D} = -k\rho v\mathbf{v}$$

where $v = |\mathbf{v}|$ is the speed.

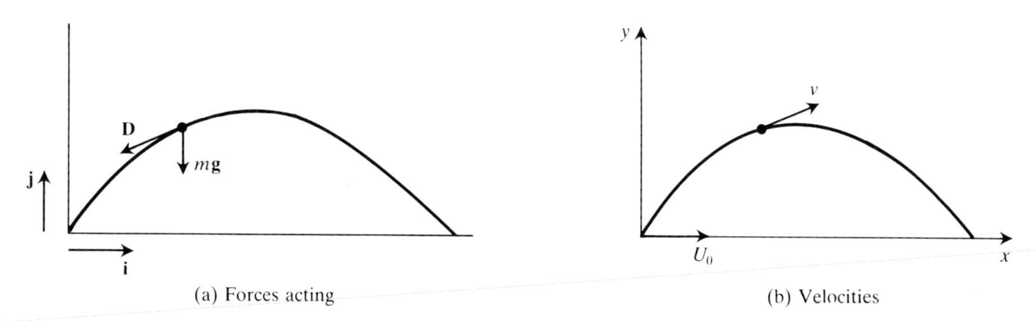

(a) Forces acting

(b) Velocities

Fig. 1.26

The equation of motion becomes

$$\frac{m\,\mathrm{d}\mathbf{v}}{\mathrm{d}t} = m\mathbf{g} - k\rho v\mathbf{v},$$

i.e.

$$\frac{\mathrm{d}\mathbf{v}}{\mathrm{d}t} = \mathbf{g} - Kv\mathbf{v} \tag{1.7.2}$$

where $K = k\rho/m$. Now the velocity vector is defined by

$$\mathbf{v} = \frac{\mathrm{d}\mathbf{r}}{\mathrm{d}t}$$

where $\mathbf{r} = x\mathbf{i} + y\mathbf{j}$ is the position vector of the particle.

Thus we can write

$$\mathbf{v} = \dot{x}\mathbf{i} + \dot{y}\mathbf{j}$$

and

$$\frac{\mathrm{d}\mathbf{v}}{\mathrm{d}t} = \ddot{x}\mathbf{i} + \ddot{y}\mathbf{j}.$$

Substituting in the equation of motion (1.7.2) gives

$$\ddot{x}\mathbf{i} + \ddot{y}\mathbf{j} = -g\mathbf{j} - Kv(\dot{x}\mathbf{i} + \dot{y}\mathbf{j}).$$

Taking components gives

$$\ddot{x} = -Kv\dot{x}$$
$$\ddot{y} = -g - Kv\dot{y} \tag{1.7.3}$$

and, since $v = |\mathbf{v}| = (\dot{x}^2 + \dot{y}^2)^{\frac{1}{2}}$, we have

$$\ddot{x} = -K(\dot{x}^2 + \dot{y}^2)^{\frac{1}{2}}\dot{x}$$
$$\ddot{y} = -g - K(\dot{x}^2 + \dot{y}^2)^{\frac{1}{2}}\dot{y}.$$

Unfortunately these equations cannot be solved exactly, but we can make progress by noting that the horizontal speed $\dot{x}$ is very much larger than $\dot{y}$, the vertical speed. Thus in (1.7.3) we can approximate

$$v = (\dot{x}^2 + \dot{y}^2)^{\frac{1}{2}} \simeq \dot{x}.$$

This gives

$$\ddot{x} = -K\dot{x}^2 \tag{1.7.4}$$

$$\ddot{y} = -g - K\dot{x}\dot{y}. \tag{1.7.5}$$

Now we can solve (1.7.4) by substituting $\dot{x} = p$, giving

$$\frac{\mathrm{d}p}{\mathrm{d}t} = Kp^2$$

and, separating variables,

$$\frac{1}{p^2}\,\mathrm{d}p = -K\,\mathrm{d}t$$

$$\frac{-1}{p} = -Kt + A$$

giving

$$\dot{x} = p = \frac{1}{Kt - A}.$$

If at $t = 0$, the initial horizontal speed is U_0 (Fig. 1.26(b)), then

$$U_0 = \frac{-1}{A}$$

giving

$$\dot{x} = \frac{1}{Kt + 1/U_0} = \frac{U_0}{1 + U_0Kt}.$$

Integrating again,

$$x = \frac{1}{K}\ln(1 + U_0Kt) + B$$

and, since at $t = 0, x = 0$, we have $B = 0$. Hence

$$x = \frac{1}{K}\ln(1 + U_0Kt) \qquad (1.7.6)$$

This formula gives us the *range* but still in terms of the time t. Expanding $\ln(1 + U_0Kt)$ in (1.7.6) gives

$$x = \frac{1}{K}(U_0Kt - \tfrac{1}{2}U_0^2K^2t^2 + \ldots),$$

i.e.

$$x = U_0t - \tfrac{1}{2}U_0^2t^2K + \ldots$$

Since K is small, the third and subsequent terms in the expression may be ignored. Hence an approximation to the difference in distances jumped at two stadiums with different air densities ρ_1 and ρ_2 will be given by

$$x_1 - x_2 = (U_0t)^2K(\rho_2 - \rho_1)/2m \qquad (1.7.7)$$

As expected if $\rho_2 < \rho_1$, then $x_2 > x_1$. Estimates for the other values in (1.7.7) are given by

$$U_0 = 10\,\text{m s}^{-1}, \quad t = 1\,\text{s}, \quad m = 80\,\text{kg}, \quad K = 0.17\,\text{m}^2$$

and

$$\rho_2 = 0.984\,\text{kg m}^{-3} \quad \text{at Mexico City}$$

$$\rho_1 = 1.225\,\text{kg m}^{-3} \quad \text{at sea level;}$$

thus

$$x_1 - x_2 = 0.026\,\text{m}\ (\approx 1\ \text{inch})$$

and this is nowhere near enough to explain the Beamon jump.

Thus we are forced to conclude that the jump was indeed miraculous!

• Problem

Consider the effect of reduced air density on the high jump.

2. Driving

2.1. Design of speed bumps

Kinematics

A common way of discouraging speeding on a housing estate is to place 'sleeping policemen' or speed bumps on the road. Unless the driver is willing to risk losing his exhaust system, he is forced to slow down and so become less of a road hazard.

Let us assume that the entry road to the estate is long and straight, and that we want to restrict speed to about 20 m.p.h. so that drivers can stop if a young child runs out into the road. A 'model B standard speed bump'—as produced by the road mending team—will slow cars down to about 5 m.p.h. The problem for the highway engineer is to decide how far apart these bumps should be placed so that they do the job.

• Solution

To determine how far apart to place the speed bumps, we must decide what sort of vehicle we are trying to slow down—clearly there is a big difference between a heavily laden lorry and a sports car.

Figures 2.1 and 2.2 show how the vehicle comes over a bump at speed v_0 (about 5 m.p.h.), accelerates up to a maximum speed v_1 (about 20 m.p.h.), and then decelerates to v_0 for the next bump. We want to choose the separation between the bumps, D, so that it will restrict *all* drivers to a maximum speed v_1—it is no good choosing D so that it will restrict the average driver but will allow the 'maniac' to do what he wants. With this in mind, we can say that our aim is to restrict the 'maniac'—someone who accelerates very hard from v_0 to v_1, and then brakes very hard from v_1 to v_0. Acceleration and braking figures for sports cars are probably a good guide.

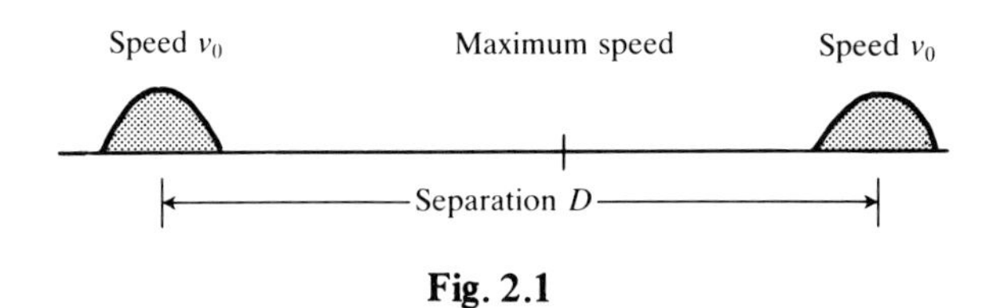

Fig. 2.1

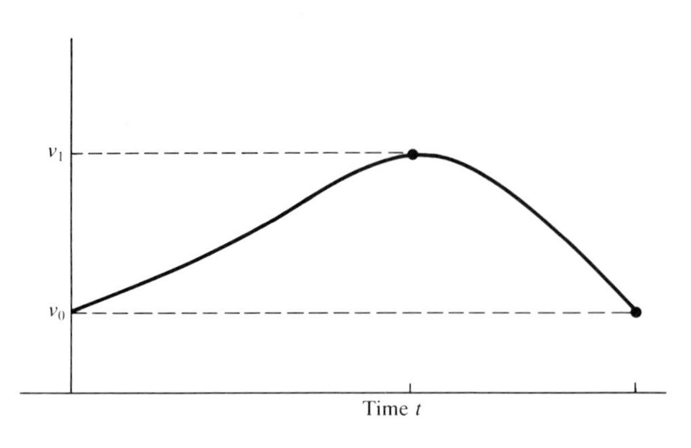

Fig. 2.2

It would be possible, at this point, to collect real velocity–time data from tests on sports cars, and use them (somehow) to work out the bump separation. However, we are after a qualitative answer—not a value for D correct to three significant figures—so it is probably sufficient to assume constant acceleration and deceleration. Then the wobbly lines in Fig. 2.2 are replaced by straight lines of gradient a (acceleration) and $-b$ (deceleration).

Using the standard formulae for motion with constant acceleration, we have

$$v_1^2 = v_0^2 + 2ad_1$$

and

$$v_0^2 = v_1^2 - 2bd_2.$$

Here d_1 and d_2 are the distance from the bump to the point of fastest speed and the distance from the point of fastest speed to the bump, respectively, and so

$$D = d_1 + d_2.$$

Rearranging this gives

$$D = \frac{v_1^2 - v_0^2}{2}\left(\frac{1}{a} + \frac{1}{b}\right). \tag{2.1.1}$$

We know that v_0 is roughly 5 m.p.h. and v_1 is roughly 20 m.p.h., so to find D and complete the problem we need to estimate a and b.

To estimate a we can consider a sports car accelerating from 0 to 60 m.p.h. in about 10 seconds. If we assume (very roughly) that this is constant acceleration throughout, then

$$60 \times \frac{5280}{3600} = 0 \rightarrow 10a$$

or

$$a = 8.8 \text{ ft/s}^2.$$

Estimating b is slightly harder, but Chapter 6 on 'Emergency stops' quotes the Highway Code figure for $b = \frac{2}{3}g$ (where g is the gravity constant). This gives $b = 21.5 \text{ ft/s}^2$.

Putting all these values into eqn (2.1.1) gives

$$D = 64.6 \text{ ft},$$

a bump spacing equivalent to the length of about four Ford Cortinas.

Comments

Although we have arrived at a very neat answer to the original *mathematical* problem, we must now check to see if it is a sensible answer to the *real* problem. Would the people who live on the housing estate be happy with a speed bump every 65 feet—even if that did slow all the traffic down to a safe speed? The answer is probably 'No', and we must resign ourselves to the idea that a sensible spacing of speed bumps is not going to slow down the 'maniac'.

• Problem

To be more realistic we might turn the problem round, letting the separation D be given and using the formula to find the maximum speed v_1 which this will allow.

A reasonable value for D might be 150 ft, i.e. 10 Ford Cortinas, and so we now have

$$v_1^2 = v_0^2 + 2D\frac{ab}{a+b}$$

$$= 54 + 1873 = 1927 \text{ (in ft/s)}$$

giving

$$v_1 = 43.9 \text{ ft/s} = 30 \text{ m.p.h.}$$

Remembering that this is the speed attained by our mythical 'maniac', perhaps this answer is acceptable. It is interesting to note that the answer is fairly insensitive to v_0, the speed over the bump, and so it is certainly not worth a great deal of time designing the perfect bump.

2.2. Emergency stops!

Kinematics and Newton's second law

When travelling at a constant speed along a straight road, how quickly can a car stop in an emergency? How might the minimum overall stopping distance differ between, say, having to stop for a child who has run into the road when you are travelling at 30 m.p.h., and having to stop for a broken vehicle on a crowded motorway when you are travelling at 70 m.p.h.?

Apart from the speed of the car, how do factors such as the weight of the car, driver's reaction time, type and quality of brakes and tyres affect the minimum stopping distance?

By extracting the important features on which the stopping distance might depend, build a model so that you can predict stopping distances.

• Solution

The overall stopping distance is made up of two components:

1. The thinking distance;
2. The braking distance.

Consider each component separately.

The thinking distance

The thinking distance (d_t) is the distance travelled between the driver becoming aware of the emergency and actually applying the brakes. The time interval for this is called the reaction time. If the reaction time is r and the car is travelling at speed v during this time, then

$$d_t = rv, \qquad (2.2.1)$$

i.e. the thinking distance varies *linearly* with the speed.

The braking distance

The braking distance is the distance travelled between application of the brakes and the vehicle coming to rest. The factors affecting this distance are more complex than for the thinking distance.

First we need to know the *effective* braking force F. How this varies with speed will depend on types and quality of brakes, but a mixture of disc and drum brakes tends to produce a constant effective braking force, independent of speed.

If x is the position of the car at time t seconds after the brakes are applied, Newton's second law gives

$$m \cdot \frac{d^2x}{dt^2} = -F \qquad (2.2.2)$$

where m represents the mass of the car and (d^2x/dt^2) is the acceleration.

The value of F will depend on the design of the car and the state of wear of the brakes, but for most cars its value is approximately proportional to the weight of the car. We can then write

$$F = kmg$$

for some constant k, where g is the gravity constant.

Substituting in (2.2.2), we have

$$m \cdot \frac{d^2x}{dt^2} = -kmg. \qquad (2.2.3)$$

Dividing each side by m and integrating, we have

$$\frac{dx}{dt} = -kgt + A.$$

When $t = 0$, $(dx/dt) = v$; hence $A = v$ and

$$\frac{dx}{dt} = -kgt + v. \qquad (2.2.4)$$

Integrating again, we have

$$x = -\tfrac{1}{2}kgt^2 + vt + B.$$

When $t = 0$, $x = 0$; hence $B = 0$ and

$$x = -\tfrac{1}{2}kgt^2 + vt. \qquad (2.2.5)$$

This gives the distance travelled whilst braking in terms of time t. The total braking time (t_b) is given from eqn (2.2.4) by setting $(dx/dt) = 0$ (since the car has come to rest). We have $t_b = v/kg$. Substituting in (2.2.5) we have the total braking distance d_b given by

$$d_b = -\tfrac{1}{2}kg\left(\frac{v}{kg}\right)^2 + v\left(\frac{v}{kg}\right),$$

i.e.

$$d_b = \frac{v^2}{2kg} \qquad (2.2.6)$$

so that the braking distance varies as the *square* of the speed v.

Combining (2.2.1) and (2.2.6), the overall stopping distance is

$$d = d_t + d_b = rv + \frac{1}{2}\frac{v^2}{kg}.$$

Comment

The solution given includes two constants, r and k, which are usually found experimentally. Their values depend on the units for time, distance, and speed.

If speed v is measured in m.p.h., distance in feet, and time in seconds, then the speed in feet per second is $22/15v$ (since 60 m.p.h. = 88 ft/s); hence

$$d_t = \frac{22}{15}vr \text{ (ft)}.$$

Experimentally, it is found that a reasonable approximation is

$$r = \frac{15}{22}$$

so that a good model is

$$d_t = v. \qquad (2.2.1a)$$

In calculating the braking distance, the braking force is found to be roughly 2/3 of the weight of the car for most cars. Thus $k = \frac{2}{3}$ so

$$d_b = \frac{(22/15v)^2}{(1/2 \times 2/3 \times 32.2)} = \frac{v^2}{20}.$$

(Again d_b is in feet although the speed v is measured in m.p.h.)

Hence the overall stopping distance is given by

$$d = v + \frac{1}{20}v^2.$$

Figure 2.3 shows a graph of the solution, for $r = 15/22$ and $k = \frac{2}{3}$.

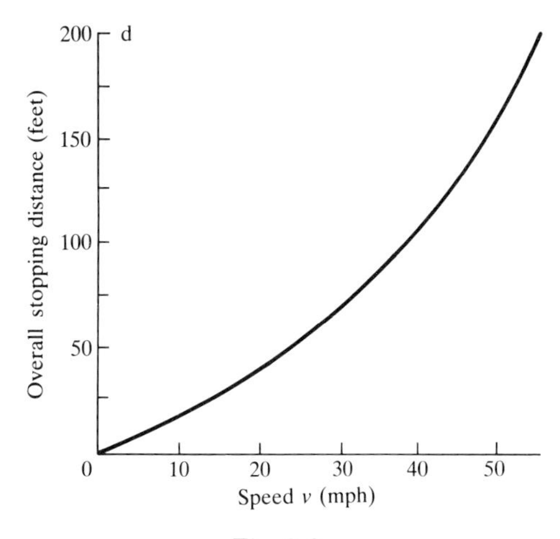

Fig. 2.3

The information on stopping distances in Table 2.1 is reproduced from the Highway Code and it will be found that the values for thinking and braking distances are those generated by our model. *Note*: The value of r could be found by the class by devising a suitable experiment of their own.

Table 2.1

Speed, v (m.p.h.)	Thinking distance, d_t (ft)	Braking distance, d_b (ft)	Overall stopping distance (ft)
30	30	45	75
50	50	125	175
70	70	245	315

Distances are based on an average family saloon.

• Problems

1. What would be the formula connecting d and v, if d were in metres and v in km/h?

2. Instead of taking $k = \frac{2}{3}$, evaluate the braking distance for $k = 0.5$ and $k = 1.0$ and examine whether variation in k between 0.5 and 1.0 significantly alters the predicted overall stopping distances.

3. It is likely that the reaction time, r seconds, used in calculating the thinking distance will vary from person to person. r was taken as the constant 15/22 in our model. If it actually varies from 0.5 to 1.0, calculate the difference between the stopping distance previously calculated and the actual range of stopping distances, for various speeds. Comment on your findings.

4. The model has been developed on the assumption of a dry road. What modification would have to be made to take account of variations in the weather? Could a 'skid factor' be incorporated? If so, where?

• Further reading

Burghes, D. N. (1981). The Humber Tunnel Authority. In *Case studies in mathematical modelling* (eds D. J. G. James and J. J. McDonald). Stanley Thornes, Cheltenham. pp. 101–9.

2.3. Designing a car braking system

Friction; Newton's second law; moments

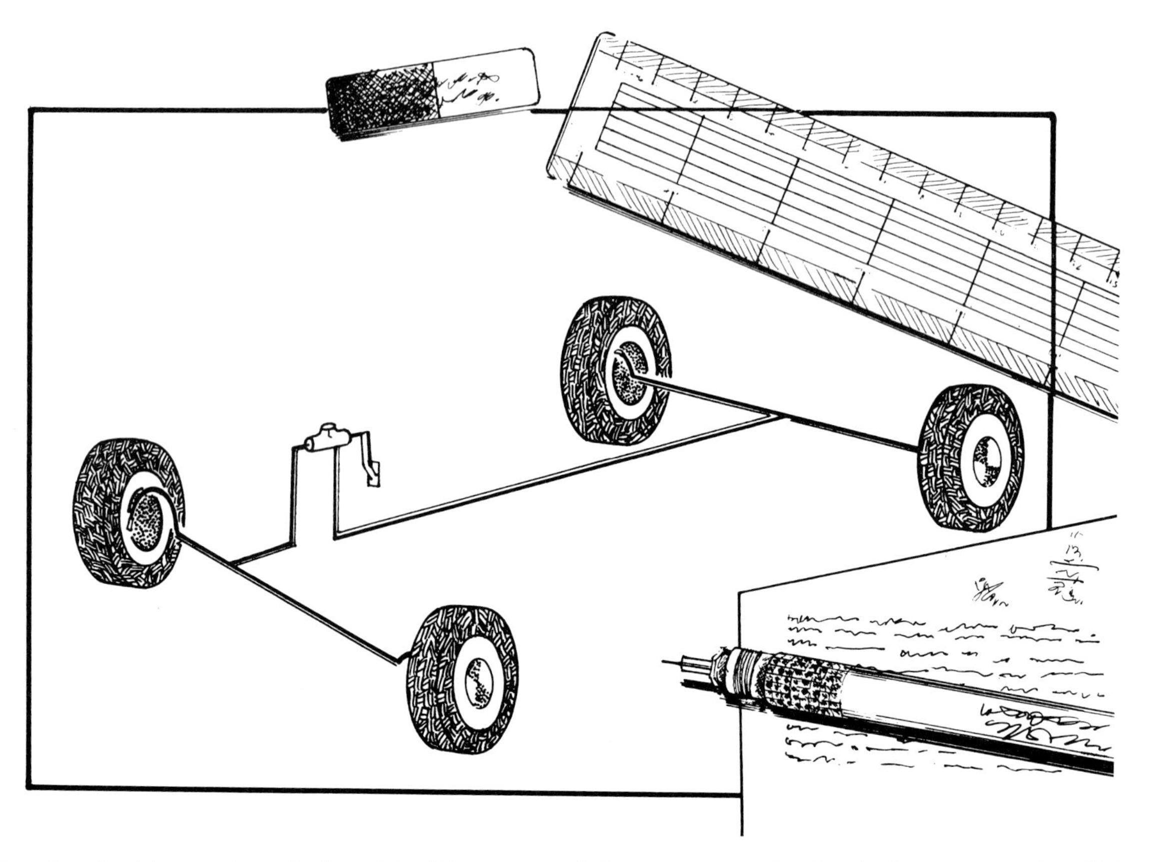

How is a braking system designed to fit a new car? At some stage in the design process, it will be necessary to design a system to fit the new model, some of whose characteristics affect the braking system. What sort of behaviour do we want it to produce?

• Solution

First of all, let us consider the car as a rigid body. The forces acting on it are as shown in Fig. 2.4, where N_F, N_R are the normal reactions on the wheels; B_F, B_R are the braking forces on the wheels; and A is the resultant of the air resistance and other resistant forces.

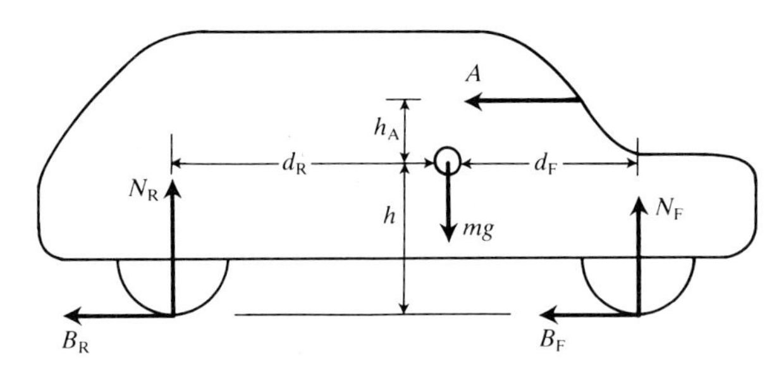

Fig. 2.4

A number of points are worth noting.

1. The position of the centre of gravity would be determined by the construction of the car, but would be arranged to coincide with what we might call the 'centre of sideways pressure'. When a crosswind is blowing this will exert a pressure on the side of the car and this can effectively be replaced by a single force acting at this centre of sideways pressure. If this point is not directly alongside the centre of gravity, the car will act like a weathervane and be difficult to control in crosswinds.
2. For similar reasons A, the 'centre of resistance', should act through the centre of gravity (i.e. $h_A = 0$).
3. In order to make the steering more effective it is probable that $d_F < d_R$ and $N_F > N_R$.

For a particular new car let us suppose that $d_F = 1$ m, $d_R = 1.5$ m, $m = 900$ kg, and $h = 0.6$ m. Newton's second law in the horizontal direction gives

$$B_F + B_R + A = 900a \tag{2.3.1}$$

where a is the retardation of the car. Newton's second law in the vertical direction gives

$$N_F + N_R - 900g = 0. \tag{2.3.2}$$

If we take moments of the forces about the position of the centre of gravity, we get

$$1.5N_R - N_F + 0.6B_F + 0.6B_R = 0. \tag{2.3.3}$$

Substituting $B_F + B_R$ from eqn (2.3.1) into (2.3.3) gives

$$1.5N_R - N_F + 0.6(900a - A) = 0. \tag{2.3.4}$$

Now we can eliminate N_F using eqns (2.3.2) and (2.3.4) so that

$$N_R = 360g - 0.24(900a - A) \tag{2.3.5}$$

and then from eqn (2.3.2) we get

$$N_F = 540g + 0.24(900a - A). \tag{2.3.6}$$

Now $B_F = \mu N_R$ and $B_F = \mu N_F$ where μ is the coefficient of dynamic friction between the tyres and road surface. The value of μ will depend on the tyres, the road surface, and other factors not under the control of the designer. μ is usually about 0.6 on wet roads and 0.75 on dry roads. Substituting in the forms of N_R and N_F, we have

$$B_R = \mu(360g - 0.24(900a - A))$$
$$B_F = \mu(540g + 0.24(900a - A)).$$

The value of A will depend on the velocity. Experimentally it is found that for slow speeds, A is approximately constant and of the order of $400N$. So

$$B_R \triangleq \mu(-216a + 3624) \tag{2.3.7}$$
$$B_F \triangleq \mu(216a + 5196) \tag{2.3.8}$$

(taking $g = 9.8 \text{ m s}^{-2}$).

The limiting values of the two braking forces (before skidding starts) are linear functions of the retardation a. This situation is illustrated by Fig. 2.5. Obviously we would like to design our braking system to maximize the retardation a, but this is limited by the frictional force of the road on the tyre; if the braking force attempts to exceed this, the wheels will lock and start to skid, thus reducing the braking force even below the limiting value. (Sliding friction is usually rather less than limiting static friction.) *So the problem is to design the brakes so that we get the maximum value of* a *before either the front or the back wheels lock.*

From Fig. 2.5 it is obvious that, for a given value of a, the limiting braking force for the rear wheels is less than that for the front wheels. This explains a number of widespread features of braking systems.

1. The use of disc brakes on front wheels are less effective than drum brakes on rear wheels. (In the days before disc brakes were invented, front wheels commonly had two hydraulic cylinders and two 'leading shoes' while rear wheels usually had only one

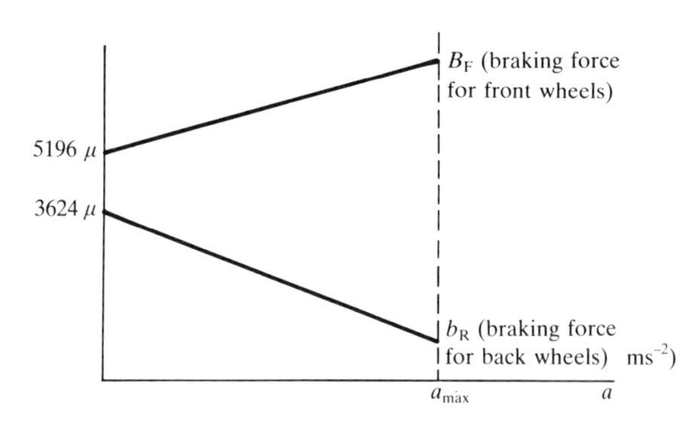

Fig. 2.5

cylinder and one 'leading' and one 'trailing' shoe which provided a lesser braking force for a given pressure on the brake pedal.)

2. The installation of pressure limiters on rear brakes, so that even when heavy pressure is applied to the brake pedal, the braking effect on the back wheels is limited to some maximum value to prevent the rear wheels skidding.

Now we consider skidding in more detail. Suppose that the total braking force is B and that a proportion of this force, p, is due to the front wheels. (The proportion due to the rear wheels is then $1-p$.) Typically p is approximately 0.7. We can write

$$B_F = pB \quad \text{and} \quad B_R = (1-p)B.$$

From eqn (2.3.1)

$$B + A = 900a$$

and then for the front brakes

$$BF = pB = p(900a - A)$$

and from eqn (2.3.8)

$$B_F = \mu(216a + 5196).$$

If the front wheels skid for a retardation of a_F, then

$$\mu(216a_F + 5196) = p(900a_F - A).$$

Solving for a_F, we get

$$a_F = \frac{Ap + 5196\mu}{900p - 216\mu}. \qquad (2.3.9)$$

A similar analysis for the rear brakes gives the acceleration for skidding, a_R,

$$a_R = \frac{A(1-p) + 3624\mu}{900(1-p) + 216\mu}. \qquad (2.3.10)$$

The optimum value of p will be the value for which the front and rear wheels start to slip simultaneously. Setting $a_F = a_R$, eqns (2.3.9) and (2.3.10) give

$$\frac{Ap + 5196\mu}{900p - 216\mu} = \frac{A(1-p) + 3624\mu}{900(1-p) + 216\mu}.$$

Solving for p, we have

$$900 \times 8820p = 5196 \times 900 + 216(8820\mu + A);$$

given values for μ and A, this equation gives the optimum value of p and then eqn (2.3.9) can be used to calculate the retardation a $(=a_F = a_R)$. For example, choosing $A = 400$ as before,

$$\text{for} \quad \mu = 0.75, \quad p = 0.780, \quad \text{and} \quad a = 7.79;$$

$$\text{for} \quad \mu = 0.65, \quad p = 0.756, \quad \text{and} \quad a = 6.81.$$

The brakes must be designed so that they deliver these proportions of the total braking force, and so that the maximum braking force provided by the brakes will produce at least the retardation given by a.

Comment

We can check this value of a for $\mu = 0.75$ by calculating the braking distance from, say, 30 m.p.h. and comparing this with the Highway Code. Converting to SI units

$$30\text{ m.p.h.} = \frac{30 \times 1609}{3600}\text{ m s}^{-1} = 13.4\text{ m s}^{-1}.$$

The braking distance in dry weather is given from the equation $v^2 - u^2 = 2as$. We have, using $a = 7.79$,

$$s = \frac{v^2}{2a} = \frac{13.4^2}{2 \times 7.79} = 11.5\text{ m},$$

which is approximately 38 feet.

The Highway Code gives 45 feet and perhaps tends to err on the side of safety! In addition, a design compromise must be made in the value of p, because the optimum value is different for dry roads and wet roads. Since braking is more of a problem in wet conditions, it seems sensible to actually use the optimum value for wet weather, $p = 0.74$ $(\mu \simeq 0.6)$. This gives a_R (which is now the limiting value) as 6.27 and the braking distance from 30 m.p.h. as 47 feet.

• Problems

1. Does the mass of the car affect the result? Repeat the calculations for a car of mass 1200 kg.

2. A rear-engined car has $d_F = 1.5$ and $d_R = 1$ instead of the other way round, all other values remaining the same. Sketch the graph of the braking forces against retardation as it would now appear. How does this affect the rest of the solution?

3. What would be the effect of lowering the centre of gravity to 0.4 m above the road?

4. How critical is the value of A? An obvious difficulty with the mathematical model as described is the use of a constant value for A, which obviously depends on the velocity. How much difference would it make to the braking distance if we ignored A completely, or if its value was four times the given value?

2.4. It's all relative!

Relative velocity

A motorist wishes to overtake a cyclist. What distance must the car travel without knocking over the cyclist?

Suppose that the cyclist moves along at about 10 m.p.h., and the car is moving at, say, 30 m.p.h. For safety, the car pulls out 10 yards before the cyclist, and moves back in 10 yards in front of him. The car is 5 yards long; the cycle is 2 yards long.

• Solution

The velocity of the car relative to the cyclist is 20 m.p.h. (30 m.p.h. − 10 m.p.h.). The problem can be treated in terms of a stationary cyclist and a car moving at 20 m.p.h. (Fig. 2.6). Since this involves only one velocity, it is an easier problem to solve.

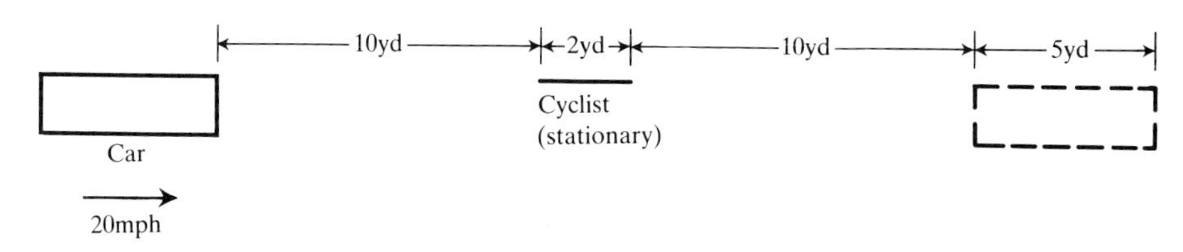

Fig. 2.6

Distance travelled by the car *relative to the cyclist*

$$= (10 + 2 + 10 + 5) \text{ yards}$$

$$= 27 \text{ yards.}$$

This gives the time taken by the car to complete the manœuvre as

$$\frac{27 \text{ yards}}{20 \text{ m.p.h.}} \quad \text{(about 2.76 seconds).}$$

In this time the actual distance travelled by the car

$$= 30 \text{ m.p.h.} \times \frac{27 \text{ yards}}{20 \text{ m.p.h.}}$$

$$= 40.5 \text{ yards}$$

or approximately 8 car lengths.

Alternatively, actual distance travelled by cyclist during the manœuvre

$$= 10 \text{ m.p.h.} \times \frac{27 \text{ yards}}{20 \text{ m.p.h.}}$$

$$= 13.5 \text{ yards.}$$

Hence actual distance travelled by the car

= distance travelled relative to the cyclist

+ actual distance travelled by cyclist

= 27 yards + 13.5 yards

= 40.5 yards, as before.

The problem can be solved without involving relative velocity as follows. Suppose that the manœuvre takes T seconds, and that in this time the car travels a distance of D yards (Fig. 2.7). Distance travelled by the cyclist

$$= D - (10 + 2 + 10 + 5) \text{ yards}$$

$$= D - 27 \text{ yards.}$$

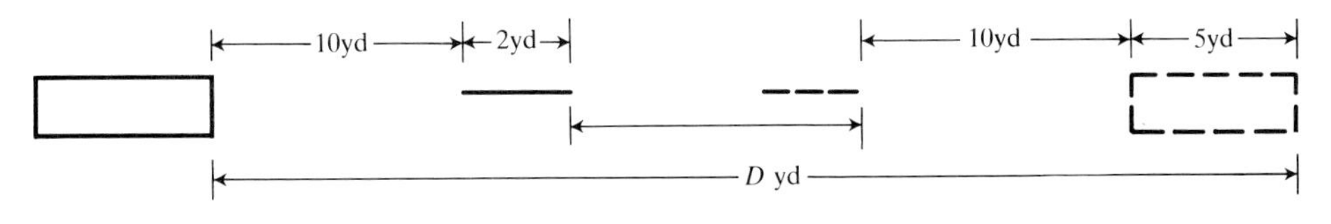

Fig. 2.7

For the car,

$$D \text{ yards} = 30 \text{ m.p.h.} \times T \text{ seconds.}$$

For the cyclist,

$$(D - 27) \text{ yards} = 10 \text{ m.p.h.} \times T \text{ seconds.}$$

Dividing,

$$\frac{D}{D-27} = 3,$$

giving $D = 40.5$ yards.

• Problem

A game at a funfair involves throwing a ball to knock down a moving target. Where does the player aim to knock down a target?

2.5. Whose fault?

Momentum; collisions

An accident between two vehicles occurs at a T-junction (Fig. 2.8). How can the police determine whose fault it is?

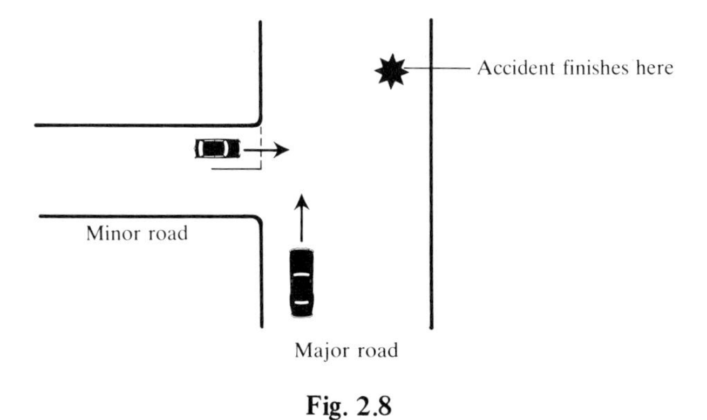

Fig. 2.8

• Solution

The police measure the skid marks and their direction and obtain results such as are shown in Fig. 2.9. We assume that the cars are locked together after the collison and use the notation

m_1 = mass of vehicle on minor road;

m_2 = mass of vehicle on major road;

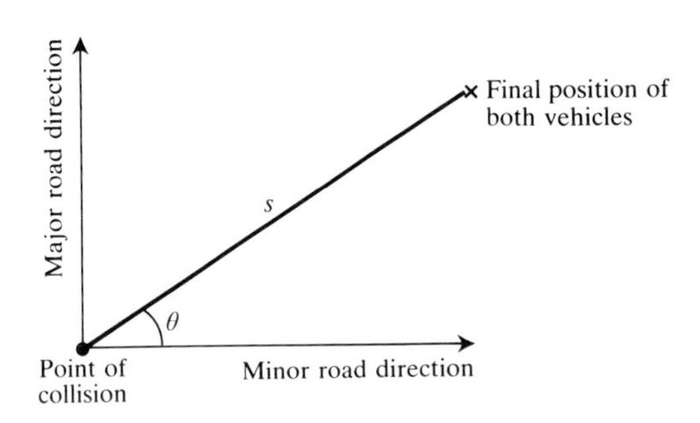

Fig. 2.9

u_1 = velocity of vehicle on minor road;

u_2 = velocity of vehicle on major road;

μ = coefficient of dynamic friction between vehicles and road;

v = velocity of both vehicles immediately after impact.

The frictional force stopping vehicles after the collision is

$$\mu(m_1 + m_2)g$$

so, equating work done with change in energy, we have

$$\mu(m_1 + m_2)gs = \tfrac{1}{2}(m_1 + m_2)v^2$$

or

$$v = (2\mu gs)^{\frac{1}{2}}.$$

Figure 2.10 shows the velocities before and after the collison. By conservation of momentum

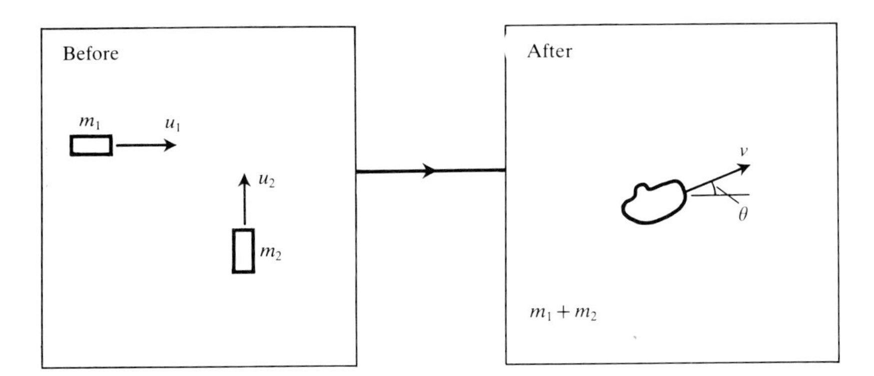

Fig. 2.10

$$m_1u_1 = (m_1 + m_2)v\cos\theta$$

$$m_2u_2 = (m_1 + m_2)v\sin\theta,$$

so

$$u_1 = \frac{(m_1 + m_2)(2\mu gs)^{\frac{1}{2}}\cos\theta}{m_1} \tag{2.5.1}$$

and

$$u_2 = \frac{(m_1 + m_2)(2\mu gs)^{\frac{1}{2}}\sin\theta}{m_2} \tag{2.5.2}$$

Equations (2.5.1) and (2.5.2) enable us to calculate the speeds of the vehicles just before impact. Hence it is possible to judge whether the vehicle on the minor road stopped at the junction or not and whether the vehicle on the major road was travelling at a reasonable speed.

Comment

This assumes that the vehicles collided at right angles and they did not rebound. It also needs a good estimate of μ to obtain satisfactory results.

• Problems

1. Evaluate u_1 and u_2 using values for m_1, m_2, μ, s, and θ that you think are reasonable.

2. Discuss the effect on u_1 and u_2 as m_1, m_2, and θ vary.

3. Write a computer program which uses the relevant data and then predicts whether the vehicle on the minor road stopped at the junction; this should take into account whether this car could have accelerated from rest to the collision velocity before collision took place.

4. Find equations similar to (2.5.1) and (2.5.2) for the velocities just before impact if the two roads met at an angle α.

5. Find equations in the case when the vehicles rebound producing two skid marks. Discuss how you would decide on a suitable value for the coefficient of restitution.

3. Further applications

3.1. Rockets
Variable mass

Rocket technology has progressed very rapidly since the launch of Sputnik 1 in 1957. The success of the space shuttle Columbia undoubtedly amazed everyone. The Apollo moonshots of the 1960s were three-stage rockets—but why should a rocket be constructed in three parts? For a rocket to leave the earth, it must overcome the gravitational pull and we will show that this necessitates constructing a rocket in stages.

• Solution

Newton's second law is often quoted as

$$\text{Force} = \text{mass} \times \text{acceleration}.$$

This, however, is a simplification which holds true only when the mass is constant. Now the mass of a rocket changes as the fuel is burnt. In fact, about 80 per cent of the mass of a rocket is fuel which is eventually burnt.

For a problem involving the coalition of two bodies (see Fig. 3.1) we go back to the form of the equation of motion that Newton gave for motion in one direction,

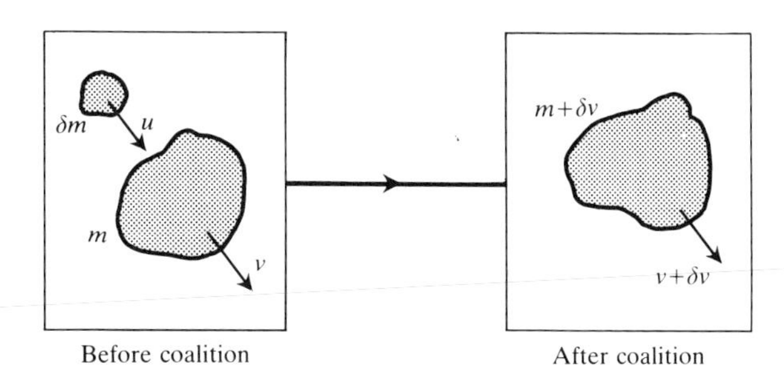

Fig. 3.1

Rockets

$$F=\frac{\mathrm{d}}{\mathrm{d}t}(mv)=\lim_{\delta t\to 0}\frac{(m+\delta m)(v+\delta v)-mv-u\delta m}{\delta t}=\lim_{\delta t\to 0}\frac{v\delta m-u\delta m+m\delta v}{\delta t}$$

$$=(v-u)\frac{\mathrm{d}m}{\mathrm{d}t}+m\frac{\mathrm{d}v}{\mathrm{d}t} \qquad (3.1.1)$$

where v is the speed of the large body at time t, F is the sum of the external forces, and u is the speed of the small body before it coalesces with the large body (this equation is derived in most mechanics textbooks).

For a rocket the fuel is the small body which is being ejected so that $u = -U$, say (see Fig. 3.2). If we let $c = v + U$, then c is the speed of the exhaust gases relative to the rocket. Equation (3.1.1) then becomes

$$F=c\frac{\mathrm{d}m}{\mathrm{d}t}+m\frac{\mathrm{d}v}{\mathrm{d}t}. \qquad (3.1.2)$$

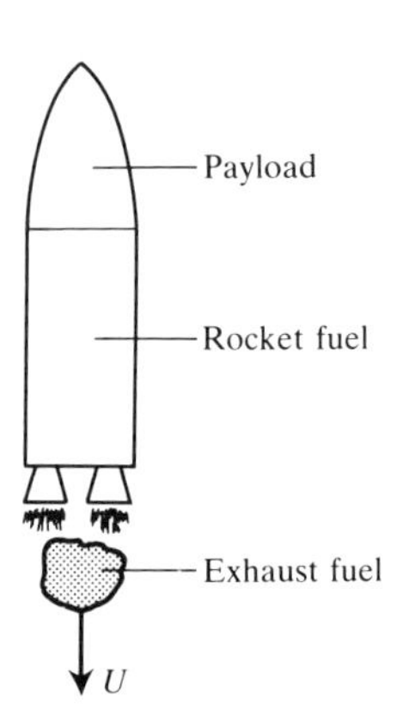

Fig. 3.2

There are two external forces acting on the rocket: (a) the gravitational force; and (b) frictional forces due to the atmosphere. Both of these forces decrease as the distance from the earth to the rocket increases. As a first attempt at solving eqn (3.1.2), all forces are ignored (i.e. $F = 0$) and eqn (3.1.2) becomes

$$m\frac{\mathrm{d}v}{\mathrm{d}t}+c\frac{\mathrm{d}m}{\mathrm{d}t}=0.$$

Further we consider c as a constant; then we can write

$$\frac{\mathrm{d}v}{\mathrm{d}m}=-\frac{c}{m}$$

and integrating we have

$$v=-c\ln m+k$$

for some constant k.

Initially, $v = 0$ and $m = P + M_0$ where P = mass of payload and the rocket casing and M_0 = initial mass of the fuel. This initial condition leads to

$$v=-c\ln\left(\frac{m}{M_0+P}\right).$$

When all the propellant has been expelled, the final mass is P and the speed is then

$$v = -c \ln\left(\frac{P}{M_0 + P}\right) = -\ c \ln\left(1 - \frac{M_0}{P + M_0}\right).$$

For example, if the mass of the fuel is 80 per cent of the total initial mass (i.e. $M_0 = 4P$) and if $c = 3\,\mathrm{km\,s^{-1}}$ then the final rocket speed will be $4.83\,\mathrm{km\,s^{-1}}$. Atmospheric drag and gravitational force will reduce this figure. The crucial question is whether this velocity is great enough to put a satellite into orbit.

For a satellite in orbit of mass P, velocity V, and distance a from the centre of the earth, the force of attraction between the satellite and the earth must equal the centrifugal force, i.e.

$$\frac{PM_E G}{a^2} = P\left(\frac{V^2}{a}\right)$$

where G is the universal gravitational constant and M_E is the mass of the earth. Solving for V gives

$$V = \frac{GM_E}{a}.$$

For example, in an orbit of height 100 km above the surface of the earth, $V = 7.8\,\mathrm{km\,s^{-1}}$ ($>4.83\,\mathrm{km\,s^{-1}}$). Thus even with this simple model it is clear why a single-stage rocket is inadequate.

• Problem

1. *The two-stage rocket*. To increase the final velocity obtained by the rocket two stages are used, as illustrated in Fig. 3.3. By considering a payload of mass $(m_2 + P)$ show that (ignoring gravitational and frictional forces) the velocity achieved at the end of the first stage is

$$-c \ln\left(1 - \frac{m_1}{m_1 + m_2 + P}\right).$$

Once the first stage has been ejected, a one-stage rocket remains. Show that the velocity may be further increased by an amount

$$-c \ln\left(1 - \frac{m_2}{m_2 + P}\right).$$

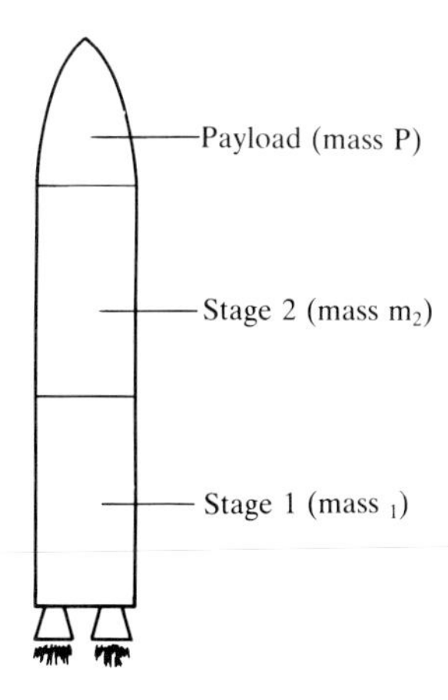

Fig. 3.3

Hence write an expression for the final velocity and evaluate it when

$$c = 3\,\text{km s}^{-2} \quad \text{and} \quad m_1 = m_2 = 50P.$$

• Further reading

Burghes, D. N. and Downs, A. M. (1975). *Modern introduction to classical mechanics and control.* Ellis Horwood, Chichester.

3.2. Communications satellites

Motion in a circle; inverse square force law; Newton's second law

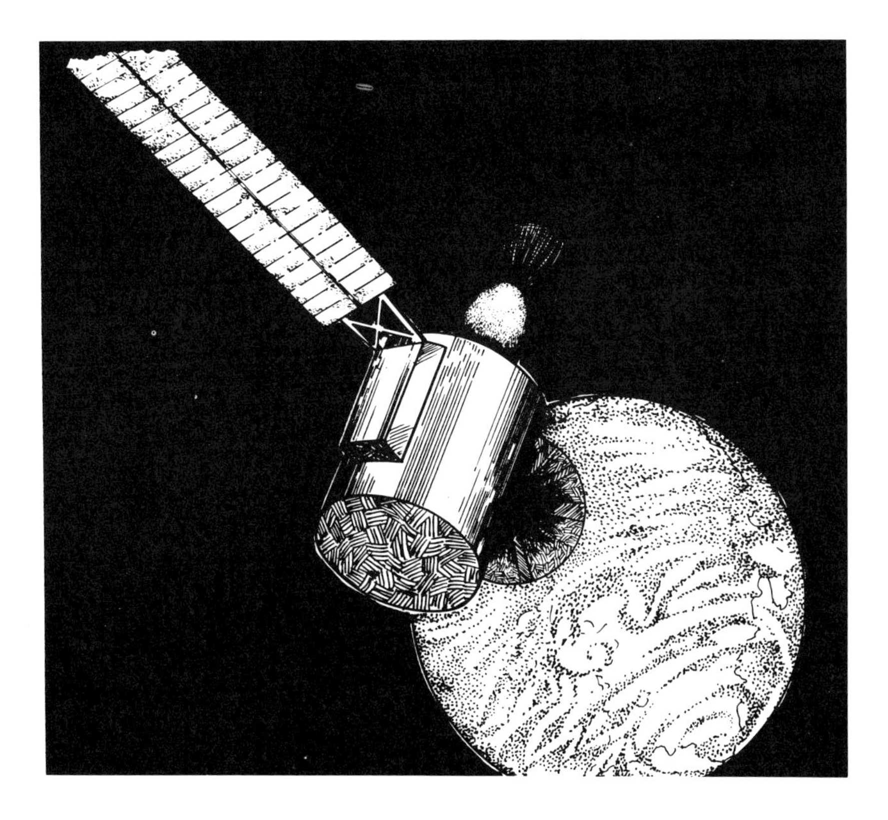

Permanent communications links between different parts of the world can be established by putting relay satellites into stationary orbits around the earth (Fig. 3.4). Although we use the word stationary, the satellite is not at rest. If the centre of the earth is considered as a fixed point then the satellite can be considered to move in a circle about this point. However, of course, the earth is spinning on its axis so that if the satellite and a point on the earth's surface have the same *angular* speed then the satellite remains above that point. To an observer the satellite appears stationary. Signals from one transmitting station on the surface of the earth are then bounced off one or more satellites until they reach their destination.

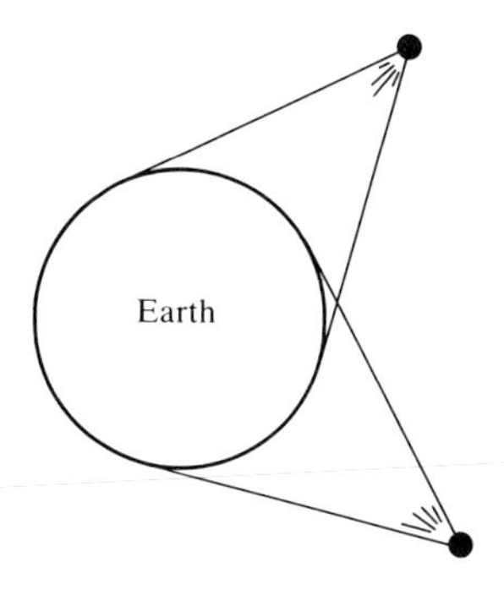

Fig. 3.4

Communications satellites

To simplify the situation we shall consider a satellite system to establish links between points on *the equator*, i.e. several satellites moving in the same circular (a further simplification) orbit above the equator.

For a single such satellite we need to consider the following questions.

1. What is the relation between the speed of a satellite and its height above the earth?
2. What then is the particular value for the height which gives a 'stationary' orbit (i.e. gives an orbiting period of 24 hours)?

Then, finally, we need to determine

3. How many such satellites are needed to link each point on the equator to every other point?

(The radius of the earth is 6.4×10^3 km approximately.)

• Solution

We consider each question separately.

1. Consider a satellite of mass m kg, moving at constant speed in a circular orbit of radius r km. Suppose that the mass of the earth is M kg and that the satellite is above the equator (see Fig. 3.5). Standard work on motion on a circle at constant speed gives the acceleration of the satellite to be along the inward radius, of magnitude v^2/r, and Newton's law of gravitation gives the force on the satellite along the inward radius to be of magnitude GmM/r^2 (where G is the universal gravitational constant).

Using Newton's second law, we have

$$\frac{GmM}{r^2} = m\left(\frac{v^2}{r}\right)$$

or

$$v = (GM/r)^{\frac{1}{2}} = \frac{(GM)^{1/2}}{r^{1/2}}, \qquad (3.2.1)$$

i.e. the speed is independent of the mass of the satellite and is inversely proportional to the square root of the radius of the orbit. It is simplest to continue to use the radius of the orbit, but if the height h above the earth is required, we need to substitute

$$r = R + h.$$

2. For one complete orbit, period T, we have

$$T = \frac{2\pi r}{v} = \frac{2\pi r^{3/2}}{(GM)^{1/2}}$$

For a *stationary* orbit, $T = 24$ hours $= 8.64 \times 10^4$ seconds, so

$$r = (8.64 \times 10^4 \times (GM)^{1/2}/2\pi)^{2/3}$$

$$= 4.22 \times 10^4 \text{km}.$$

Substituting $G = 6.67 \times 10^{-11}$ (m/kg s^2) and $M = 5.98 \times 10^{24}$ kg, we get

$$r = 4.22 \times 10^4 \text{km}$$

and then, from eqn (3.2.1),

$$v = 97.2 \text{ km s}^{-1}.$$

3. Consider Fig. 3.5 in which O is the centre of the earth. One satellite will provide communications links with the surface of the earth covered by the cone shown where OAS is a right angle. We can easily find the value of the angle α from the triangle OAS. We have

$$\cos\alpha = \frac{R}{r} = \frac{0.64 \times 10^4}{4.22 \times 10^4}.$$

Thus

$$\alpha = 81°.$$

This implies that each satellite covers about 160°. Since the total angle at the centre of the earth is 360°; two satellites will cover 320° (<360°) and three satellites will cover 480° (>360°). Thus three satellites will be sufficient to link any pair of points on the equator.

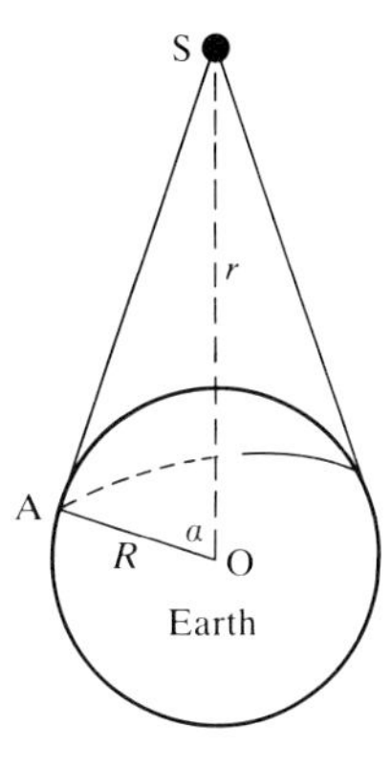

Fig. 3.5

• Problem

Kepler's third law

On a larger scale, the planets of the solar system are satellites of the sun. If we again simplify the situation by taking the planetary orbits to be circular, at constant speed, find how well does observed data of the nine planets of the sun fit the relationship between period T and radius of orbit r given by

$$T = \frac{2\pi r^{3/2}}{(GM)^{1/2}}.$$

This formula is a representation of the result found by Johann Kepler in 1619, using observations of Tycho Brahe. The latest figures available (1983) are shown in Table 3.1.

Table 3.1

Planet	Period (days)	Mean distance from sun (km $\times$ 10^6)
Mercury	88	57.9
Venus	225	108.2
Earth	365	149.6
Mars	687	227.9
Jupiter	4 329	778.3
Saturn	10 753	1427
Uranus	30 660	2870
Neptune	60 150	4497
Pluto	90 670	5907

3.3. Maximum range

Projectiles

On 4 June 1982 the British invasion forces (Fig. 3.6) had just captured the mountain position of Mount Kent (height 458 m) and the Argentinian occupation army had retreated east to the garrison at Stanley (10 km away). The biggest guns that could be moved onto the mountain top had muzzle speed of $400\,\mathrm{m\,s^{-1}}$. At what angle should they have been fired in order to hit the garrison?

After the expected destruction of the garrison, the retreating occupation army was expected to flee further eastwards towards Stanley airport. What is the effective range of attack from the mountaintop?

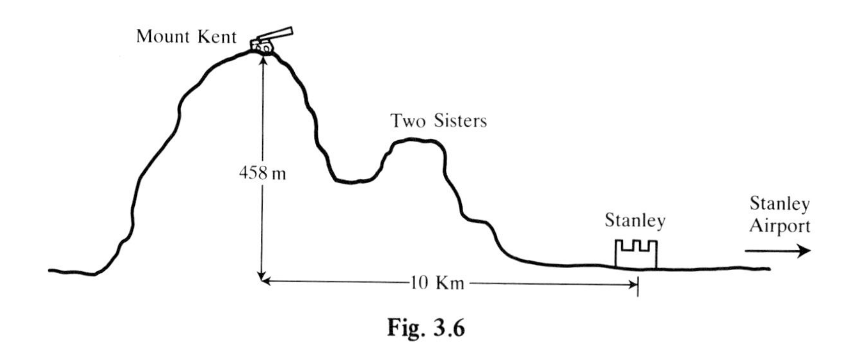

Fig. 3.6

• Solution

Let's generalize this problem to a mountain height h (km) and a garrison, horizontal distance d (km) away from the mountain. If we can solve the problem in general, we not only have a solution for this problem but all such problems.

Suppose the gun has muzzle speed v and is fired at angle α to the horizontal. Neglecting resistance (as a first approximation), the only force on the shell is gravity acting downwards. With axes as shown (Fig. 3.7), we have the equations of motion

$$\text{Horizontal component} \qquad m\frac{d^2x}{dt^2} = 0 \tag{3.3.1}$$

$$\text{Vertical component} \qquad m\frac{d^2y}{dt^2} = -mg \tag{3.3.2}$$

where m is the mass of the shell. We need to find the equation of the trajectory.

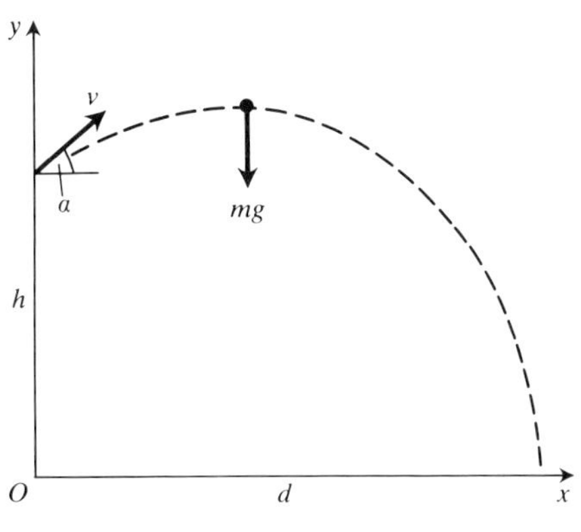

Fig. 3.7

Integrating (3.3.1) and (3.3.2) gives

$$\frac{dx}{dt} = A$$

$$\frac{dy}{dt} = -gt + B$$

At $t = 0$, initial conditions give

$$A = v\cos\alpha$$

$$B = v\sin\alpha,$$

i.e.

$$\frac{dx}{dt} = v\cos\alpha$$

$$\frac{dy}{dt} = v\sin\alpha - gt.$$

Integrating again

$$x = vt\cos\alpha + A$$

$$y = vt\sin\alpha - \tfrac{1}{2}gt^2 + B$$

and, since initially $x = 0, y = h$, we have

$$x = vt\cos\alpha$$
$$y = vt\sin\alpha - \tfrac{1}{2}gt^2 + h.$$

The path of the particle is found by eliminating t, to give

$$y = x\tan\alpha - \tfrac{1}{2}g\frac{x^2}{v^2\cos^2\alpha} + h. \qquad (3.3.3)$$

For the shell to hit the garrison at $(d, 0)$ we need

$$0 = d\tan\alpha - g\frac{d^2\sec^2\alpha}{2v^2} + h$$

and this must be solved for the angle of projection α.

This gives

$$0 = 2v^2 d\tan\alpha - gd^2(1 + \tan^2\alpha) + 2v^2 h$$

which reduces to

$$\tan^2\alpha - \frac{2v^2}{gd}\tan\alpha + 1 - \frac{2v^2 h}{gd^2} = 0.$$

Solving,

$$\tan\alpha = \frac{v^2}{gd} \pm \left(\frac{v^4}{g^2d^2} - 1 + 2\frac{v^2 h}{gd^2}\right)^{\frac{1}{2}} \qquad (3.3.4)$$

In the original problem we have $v = 400\ \mathrm{m\,s^{-1}}$, $h = 458$ m, $d = 10\,000$ m, and, taking $g = 9.8\ \mathrm{m\,s^{-2}}$, eqn (3.3.4) gives the required angle of projection as either $16°$ or $71°$. Since the time of flight on the lower trajectory will be less than the higher one, since $t = (d/v\cos\alpha)$, the suggested angle is $16°$.

The second problem requires finding the maximum range along the x-axis. Defining this by R, (3.3.3) gives

$$0 = R\tan\alpha - \frac{gR^2}{2v^2\cos^2\alpha} + h. \qquad (3.3.5)$$

Now for maximum R, we must have $\mathrm{d}R/\mathrm{d}\alpha = 0$. Differentiating through the expression with respect to α gives

$$0 = \frac{\mathrm{d}R}{\mathrm{d}\alpha}\tan\alpha + R\sec^2\alpha - \frac{gR}{v^2\cos^2\alpha}\frac{\mathrm{d}R}{\mathrm{d}\alpha} - g\frac{R^2}{v^2}\sec^2\alpha\tan\alpha$$

and putting $\mathrm{d}R/\mathrm{d}\alpha = 0$ results in

$$R\sec^2\alpha = g\frac{R^2}{v^2}\sec^2\alpha\tan\alpha,$$

i.e.

$$\tan\alpha = v^2/gR. \qquad (3.3.6)$$

Substituting back in (3.3.5) gives

$$0 = \frac{v^2}{g} - \frac{gR^2}{2v^2}\left(1 + \frac{v^4}{g^2R^2}\right) + h$$

and solving for R results in

$$R^2 = \frac{v^4}{g^2}\left(1 + \frac{2gh}{v^2}\right)$$

or

$$R = \frac{v^2}{g}\sqrt{\left(1 + \frac{2gh}{v^2}\right)} \tag{3.3.7}$$

For the original problem, we take

$$v = 400\ \mathrm{m\,s^{-1}}, \quad h = 458\ \mathrm{m}$$

and, with $g = 9.8\ \mathrm{m\,s^{-2}}$, we have $R = 16.8$ km (from eqn (3.3.6) the angle of projection is 44.2°). In fact, Stanley Airport is a further 4 km east of Stanley, and so the Argentinian forces would have remained in range of the British guns.

• Problems

1. Find the angle of projection of the shells if the muzzle speed is

$$\text{(i) } 600\ \mathrm{m\,s^{-1}}; \quad \text{(ii) } 300\ \mathrm{m\,s^{-1}}; \quad \text{(iii) } 100\ \mathrm{m\,s^{-1}}.$$

What is the explanation of the result in (iii)?

2. Figure 3.8 shows the path of the centre of gravity of a long jumper. With the data given, find the length of the jump.

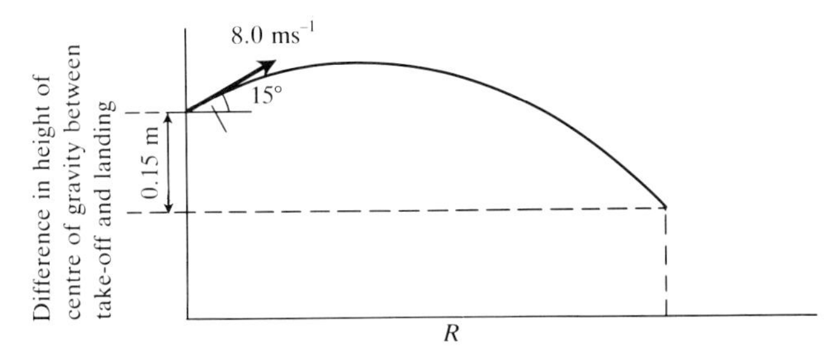

Fig. 3.8

3.4. Waterwheels

Work and energy

At present, the power used in our homes and in industry is derived almost exclusively from fossil fuels. The problem of supplying power when these fuels are exhausted is already urgent. In the long run, the solution may lie in nuclear power, but meanwhile a good deal of thought is being given to the alternatives offered by moving water (tides, waves, and streams) or wind, or the direct heat of the sun. Some of these will make a significant contribution in bridging the 'energy gap' and hydroelectric power is already important in some parts of the world, including Wales and Scotland. Hydroelectric power is obtained from moving water using the turbine, a development of the waterwheel which was for many centuries in Europe the main source of mechanical power. Waterwheels were still used in this country in large numbers at the beginning of the industrial revolution, chiefly in the manufacture of woollen goods and for milling grain. Is there a case for reintroduction of the waterwheel in installations where a modest power supply (say, up to 100 kW) is needed?

There are three main types of wheel (see Fig. 3.9) of which we shall consider here only two: the undershot and the overshot. Of these, the simpler mechanically is the undershot wheel. This has a set of equally spaced blades, or paddles, fixed radially in its perimeter. The blades dip into a moving stream of water, usually in an artificially constructed channel, and the wheel obtains its power from the momentum of the water. The overshot wheel is more commonly found in hilly areas. Its power is derived from the weight of water carried in its buckets from the head race, which delivers the water at the top, to the tail race, at the base of the wheel. Let us consider the undershot and overshot wheels separately.

THE OVERSHOT WHEEL

In effect, this device (Fig. 3.10) saves the kinetic energy usually dissipated in a waterfall. The buckets may simply be shaped blades or they may be hinged. However carefully designed, the buckets will spill some of their contents before reaching the base of the wheel. For the purposes of our calculation, though, we shall assume that all the water flowing from the head race is

Water wheels

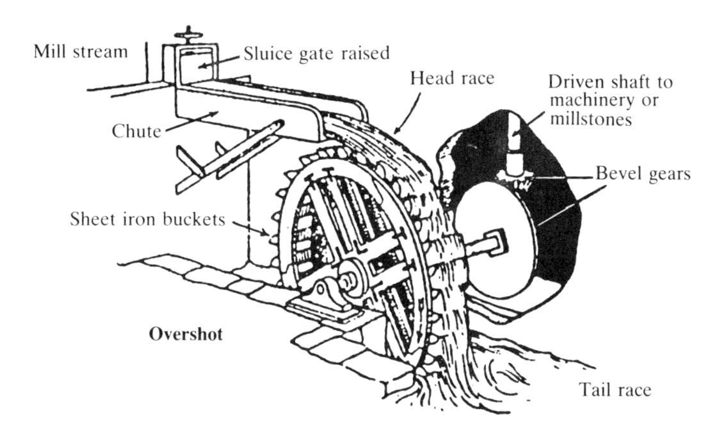

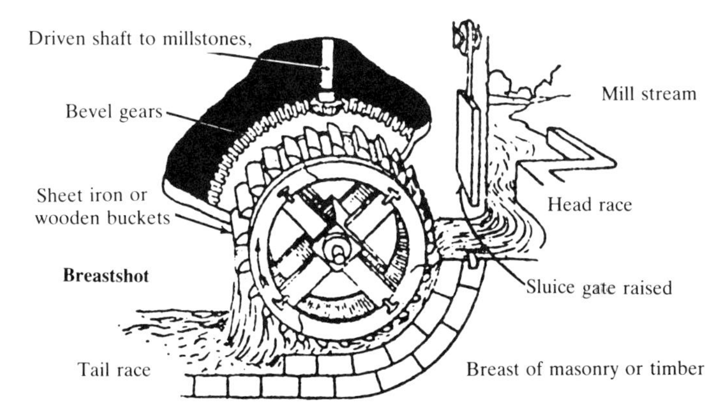

Fig. 3.9

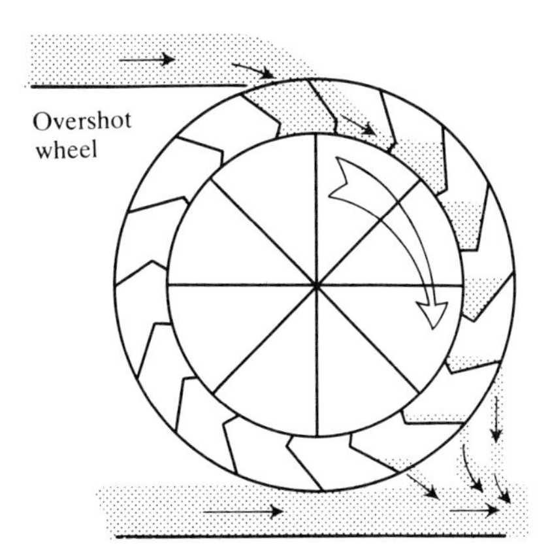

Fig. 3.10

delivered with zero kinetic energy into the tail race. Suppose that the volume flow from the head race is $V\,\mathrm{m^3\,s^{-1}}$ and the depth of the fall is h metres. If the efficiency of the wheel is η, what power does it deliver?

• Solution

The water in the head race has both kinetic energy and potential energy (because of the difference in height between the top and bottom of the wheel). Since the kinetic energy is the result of an almost horizontal flow whose effect is simply to deliver the water into buckets, we shall assume that it does no useful work. When it emerges from the wheel the water has zero energy, so the whole of its initial potential energy has been given up to the wheel.

Water has a density of 1 gram per cubic centimetre or $10^3\,\mathrm{kg\,m^{-3}}$, so the mass of water falling each second is $10^3 V$ kg. Hence the force exerted by gravity is $10^3\,Vg$ newtons and the power supplied is $10^3\,Vgh\ \mathrm{N\,m\,s^{-1}}$ (watts). Of this, a proportion η is useful. Thus the answer is

$$10^3\,Vgh\eta \text{ watts, or } \frac{10^3\,Vgh\eta}{746} \text{ horse power.}$$

Example

If the volume flow is $10\,\mathrm{m^3\,s^{-1}}$, the fall is 5 m, and the efficiency is 70 per cent, then the power is

$$10^3 \times 10 \times 9.81 \times 5 \times 0.7 \text{ watts}$$
$$= 340\,\text{kW (or 460 h.p.).}$$

THE UNDERSHOT WHEEL

Though this device (Fig. 3.11) is mechanically simpler, its theory is more complicated. The water in the channel moves with speed $U\,\mathrm{m\,s^{-1}}$ and drives the wheel so that the mid-point of a blade moves with speed $u\,\mathrm{m\,s^{-1}}$. A detailed analysis would involve integration of the components of force acting perpendicular to each blade as the blades rotate. In practice, a reasonably accurate

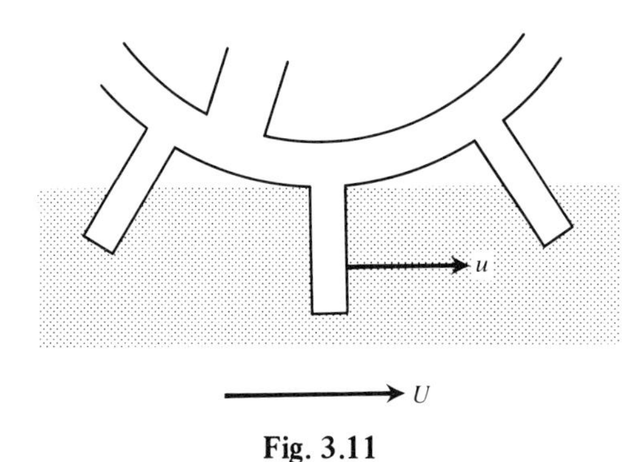

Fig. 3.11

result is obtained by assuming that at each instant the full force of the flow is directed at right angles to a single blade. We shall assume that the drag of the walls of the channel may be neglected (i.e. that the force is uniformly distributed over the whole area of the blade). Finally, effects of turbulence and water rising up the blade or being carried on the blade will be neglected.

Suppose the blade area immersed is $A\ \mathrm{m^2}$ and the efficiency of the wheel is η. Once again, the problem is to determine the power delivered by the wheel. This will vary with u, so we shall calculate its maximum value.

• Solution

Before impact the speed of the water is U. After impact we shall take it to be u. This is a reasonable assumption if the wheel works in a channel, so that water cannot escape at the side of the blade.

The mass of water impinging on the blade each second is $10^3 A(U-u)$ kg (since $U-u$ is the speed in m s^{-1} of the water relative to the blade). The speed of the water falls from U to u, so that the rate of loss of momentum is

$$F = 10^3 A(U-u)^2 \text{ kg m s}^{-2}.$$

This is the action of the blade on the water, and is accompanied by an equal and opposite reaction of the water on the blade. We note that rate of change of momentum is force.

Taking $u \geqslant 0$ the force is a maximum when $u = 0$ and decreases to zero when $u = U$. (If the wheel was immersed in the water and doing no work it would tend to speed up till $u = U$. In practice, if no work is required for a long period the water flow is diverted from the channel using a sluice gate.) Thus the maximum force is

$$F_{\mathrm{M}} = 10^3 A\, U^2 \quad \text{(in newtons)}.$$

The power supplied is

$$\begin{aligned} P &= \text{rate of work of water on wheel} \\ &= \text{force} \times \text{distance moved per second} \\ P &= 10^3 A(U-u)u^2\eta \text{ N m s}^{-1}. \end{aligned} \tag{3.4.1}$$

P varies with u and will be a maximum when $\mathrm{d}P/\mathrm{d}u = 0$. The graph of P against u is part of a cubic curve (Fig. 3.12).

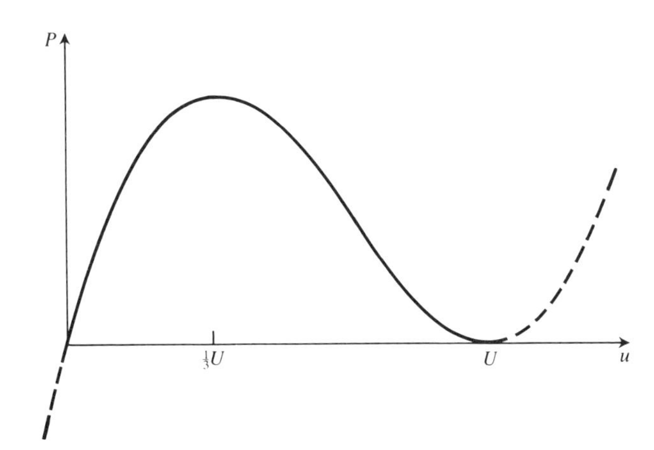

Fig. 3.12

Differentiating eqn (3.4.1),

$$\frac{10^{-3}}{A}\frac{\mathrm{d}P}{\mathrm{d}u} = (U-u)^2 - 2u(U-u)$$

$$= (U-u)(U-3u);$$

and $(\mathrm{d}P/\mathrm{d}u) = 0$ when $u = \frac{1}{3}U$ ($u = U$ gives zero power, the minimum).

We see that the maximum power $\hat{P}$ is supplied when $u = U/3$ (and minimum power when $u = U$). Thus maximum power is

$$\hat{P} = 10^3 A(U - U/3)^2 \cdot \tfrac{1}{3} U\eta$$

$$= \frac{4}{27} A\, U^3 \eta \times 10^3 \text{ watts.}$$

Example

An undershot wheel has paddles 2 m long, immersed in water to a depth of 50 cm. The radius of the wheel (say, distance from axle to midpoint of paddle) is 4 m. The speed of the water in the channel is initially 10 m s^{-1} and it is slowed to 5 m s^{-1} by the wheel. Taking the efficiency of the wheel to be 0.6 find

(i) The torque exerted by the water on the axle;

(ii) The power of the wheel;

(iii) The maximum power of which the wheel is capable.

(i) The torque is the moment exerted by the force on a paddle, i.e. FR newton metres, where R metres is the radius. In this case, $F = 10^3 \times 1 \times (10 - 5)^2$ and $R = 4$. So the torque is 10^5 N m.

(ii) Using the formula $P = 10^3 A(U - u)u^2\eta$, we find the power is $10^3 \times 1 \times 5^2 \times 5 \times 0.6$ watts, i.e. 75 kW or just over 100 h.p.

(iii) The maximum power is given by

$$P = 10^3 \times \frac{4}{27} A\, U^3 \eta,$$

and is

$$10^3 \times \frac{4}{27} \times 1 \times 10^3 \times 0.6 \text{ watts,}$$

i.e. 89 kW or just under 120 h.p.

Comments

The value of η in each case would be in the range 60–75 per cent. The notion of efficiency in the undershot case is capable of two interpretations. Here we have taken it as the mechanical efficiency of the wheel itself. We might take it instead as being the ratio

$$\frac{\text{Work output}}{\text{Kinetic energy}}$$

for a given mass of water (say, the mass impinging on the wheel each second). Since the kinetic energy is the work done in slowing the water to zero speed and for maximum power we only slow it to one-third its initial speed, we are using only a part of the power of the water. The mass of water impinging in one second is $10^3 A U$ kg and so its kinetic energy is $\frac{1}{2} \times 10^3 A U^3$ joules. Since the energy supplied to the wheel each second is, at maximum, $10^3 \times 4/27\, A U^3$ joules. we see that only 8/27 or less than 30 per cent of the kinetic energy is available to the wheel.

Historically, the earliest recorded use of waterwheels was for irrigation purposes in the Middle East, in the second century BC. The overshot wheel is thought to have been developed by the Romans and its use was widespread and general in Europe up to the end of the eighteenth century when steam power was exploited. The old millers' rule of thumb for maximum power of an undershot wheel was to have the paddle moving at a speed between 1/3 and 1/2 that of the water, which agrees well with the figure obtained in our model.

• Problems

1. An undershot wheel has paddles which are 4 m wide and dip 0.4 m into a stream whose flow is 3 m s^{-1}. Find the force on the paddle blades when vertical and the horsepower developed by the wheel when the paddles are moving at 0.8 m s^{-1}.

What is the highest power which the wheel could generate?

2. The diameter of the wheel does not come into the calculations for the power generated by an undershot wheel. Under what circumstances might a millwright decide to use (i) a small-diameter wheel; (ii) a large-diameter wheel?

3. Draw a graph of the power output P of an undershot wheel as it changes with its paddle speed u and find the range of values of u for which the wheel generates at least 75 per cent of its peak power.

4. A tidal mill pond whose area is 3×10^5 m^2 is filled to a depth of 2 m at every tide. A large undershot wheel is to be operated by the water flowing out of this pond along a channel as the tide recedes. Assuming that there is a sufficient head of water for the water to flow along a channel at 4 m s^{-1} for 4 hours each tide, decide on the dimensions of the channel and waterwheel to generate the maximum power possible.

5. Water flows along a leat (head-race) at a speed of 1.5 m s^{-1} to the top of an overshot waterwheel whose diameter is 8 m. If the width of the leat is 2 m and the water flows at a depth of 0.1 m, estimate the likely power output of the wheel if it is 63 per cent efficient.

6. The Devon Great Consols Mine at its peak of activity had 33 overshot wheels in operation and obtained the water to drive many of these by building a weir across the River Tamar at Latchley from where a leat over 3 km long and 5.6 m wide supplied a steady flow of water. One of these wheels was 16 m in diameter and 4 m wide and developed up to 130 h.p. Assuming the wheel to be 60 per cent efficient, estimate the flow of water it required in cubic metres per second.

7. On hilly country where a large fall of water can easily be arranged, waterwheels are particularly applicable. The Wheal Friendship mine on the edge of Dartmoor worked for 130 years and made very extensive use of water power by bringing two leats from the River Tavy producing a fall of 160 m to drive 17 overshot wheels. If the average volume flow of water along these leats was 0.5 m^3 s^{-1}, what horsepower would be potentially available from this fall of water, assuming 50 per cent efficiency? The annual rainfall on this part of Dartmoor is in excess of 2 m. What catchment area would be equivalent to the above volume flow?

• Further reading

Booker, F. (1967). *Industrial archaeology of the Tamar Valley*. David and Charles, Newton Abbot.

Cossons, N. (1975). *The BP book of industrial archaeology*. David and Charles, Newton Abbot.

Rees, A. (1972). *Rees's manufacturing industry (1819–20)* (Ed. N. Cossons), Vol. 5. David and Charles, Newton Abbot.

Schools Council (1975). *Industrial archaeology of watermills and waterpower*, Project Technology Handbook 11. Heinemann Education Books, London.

Woodall, F. D. (1975). *Steam engines and waterwheels*. Moorland Publishing Company, Ashbourne, Derbyshire.

3.5. Raindrops

Rain is formed by condensation of water vapour some distance above the ground to form clouds, where droplets aggregate to make drops of sufficient size to fall. These drops increase in size as they pass through the cloud and eventually reach the ground after suffering more or less from evaporation as they fall through clear air.

Our weather being what it is, we have an infinite variety of kinds of rain. The clouds may be turbulent, with strong upwards currents within them, or they may be in a fairly steady state. They may be quite high or they may extend to the ground—especially if you happen to be on a hill. After they have formed, the raindrops may fall through still air or be driven by the wind. The air may be warm or cold, humid or dry. Because of all these variables we experience every sort of rain from fine drizzle to giant thunderstorm drops, coming at us with a wide range of speeds and inclinations.

As a first effort at modelling a particular sort of rain, suppose that there is a horizontal layer of cloud of uniform depth d, whose base is at height h above horizontal ground (Fig. 3.13). Conditions everywhere are calm, with no strong currents. Suppose that rain forms in the upper cloud, starting to fall as very small drops. Suppose that these increase in size as they pass through the cloud. Finally, suppose that these drops suffer no loss by evaporation as they fall through the clear air.

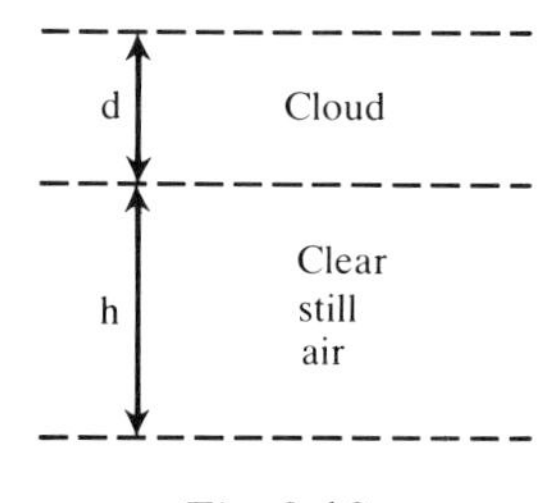

Fig. 3.13

• Problems

1. Suppose that the rate of increase of mass of a drop as it falls through the cloud is proportional to the product of its surface area and its velocity. Show that its radius increases linearly with the distance it has fallen.

2. In its fall through the cloud, suppose that the drop is subject to a resistance proportional to the product of its surface area and its velocity. Show that if s is the distance fallen and v the velocity, then

$$v\frac{dv}{ds} = g - \frac{cv}{s} - \frac{3v^2}{s}$$

where c is a constant.

3. Assuming c to be very small (so that the inconvenient term cv/s may be ignored!) show that after a long fall through cloud the acceleration of a drop would tend to $g/7$.

4. If the velocity of the raindrops on leaving the cloud is u and they are subject to a resistance Kv^2 per unit mass when their velocity is v, find with what velocity they reach the ground.

• Suggested solutions

1. When the radius of the drop is r, assuming it to be spherical, its mass is Ar^3 ($A = 4/3\pi\rho$, where ρ is the density), and its surface area is $4\pi r^2$. Hence, from the given proportionality,

$$\frac{d}{dt}(Ar^3) = B\cdot 4\pi r^2 v$$

where B is a constant.

Now $v = ds/dt$, so

$$3Ar^2\frac{dr}{dt} = 4B\pi r^2\frac{ds}{dt}.$$

Thus

$$\frac{dr}{dt} = m\frac{ds}{dt}$$

where $m = 4\pi B/3A$ is a constant,

$$r = ms + c$$

and, by selecting a suitable origin, the constant c may be made zero. In our case, $c = 0$ since we have assumed negligible radius at the start of the fall from the top of the cloud.

2. The resistance to motion is Dr^2v, where D is a constant, so the forces acting on the drop with radius r and velocity v are as shown in Fig. 3.14. Equating rate of change of momentum to force we have

$$\frac{d}{dt}(Ar^3v) = Ar^3g - Dr^2v.$$

Since $r \propto s$, cancelling by various constants,

$$\frac{d}{dt}(s^3v) = s^3g - cs^2v$$

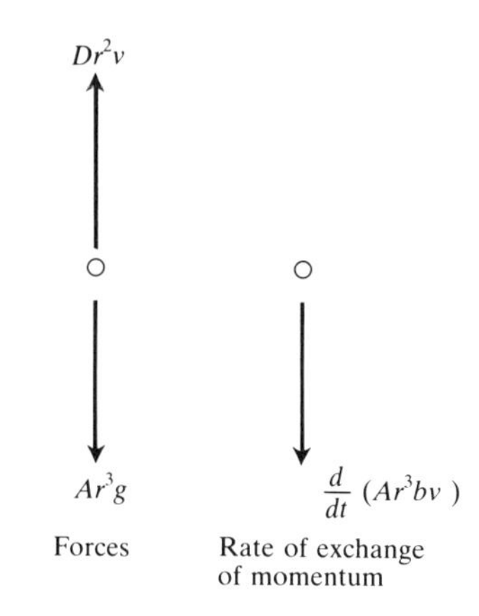

Fig. 3.14

where c is a constant.

$$3s^2\left(\frac{ds}{dt}\right)^2 + s^3\frac{dv}{dt} = s^3g - cs^2v$$

$$\frac{dv}{dt} = g - \frac{cv}{s} - \frac{3v^2}{s}$$

$$v\frac{dv}{ds} = g - \frac{cv}{s} - \frac{3v^2}{s}.$$

3. Ignoring the term cv/s above and putting

$$v^2 = p, \quad 2v\frac{dv}{ds} = \frac{dp}{ds},$$

the equation becomes

$$\frac{dp}{ds} = 2g - \frac{6p}{s}$$

$$\frac{dp}{ds} + \frac{6p}{s} = 2g.$$

This is a linear equation of the first order, with integrating factors s^6. Multiplying by this, we have

$$s^6\frac{dp}{ds} + 6s^5p = 2s^6g,$$

which may be written

$$\frac{d}{ds}(s^6p) = 2s^6g.$$

Integrating,

$$s^6p = \tfrac{2}{7}s^7g + E$$

where E is a constant. Then

$$p = v^2 = \frac{2gs}{7} + \frac{E}{s^6}.$$

As $s \to \infty$, $E/s^6 \to 0$, and, approximately,

$$v^2 = \frac{2gs}{7}.$$

Differentiating with respect to s,

$$2v\frac{dv}{ds} = \frac{2g}{7}$$

$$v\frac{dv}{ds} = \frac{dv}{dt} = \frac{g}{7}.$$

4. Equating the algebraic sum of forces per unit mass with acceleration Fig. 3.15),

$$v\frac{dv}{ds} = g - Kv^2$$

$$\int \frac{v\,dv}{g - Kv^2} = ds.\int$$

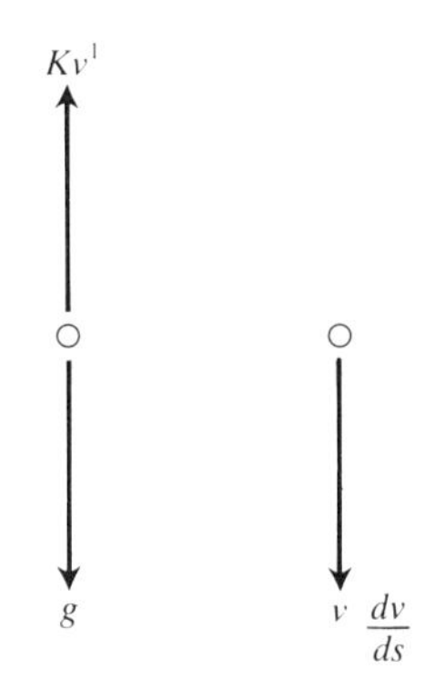

Fig. 3.15

Integrating,

$$-\frac{1}{2K}\ln(g - Kv^2) = s - \frac{1}{2K}\ln(g - Ku^2)$$

where the constant of integration is found using the initial conditions.

Then

$$\frac{1}{2K}\ln\frac{g - Kv^2}{g - Ku^2} = -s$$

$$\frac{g - Kv^2}{g - Ku^2} = e^{-2Ks}$$

$$v = \left\{\frac{g}{K} - e^{-2Ks}\left(\frac{g}{K} - u^2\right)\right\}^{\frac{1}{2}}.$$

Substituting $s = h$ gives the required velocity. We note that as $s \to \infty$, $v \to (g/K)^{\frac{1}{2}}$, the terminal velocity being independent of u.

• Problems

1. Show that if in Problem 1 we make the rate of increase of mass simply proportional to the surface area then the acceleration tended to is $g/4$ (cf. $g/7$ in Problem 3).

2. In alternative models, the resistance in clear air could be proportional to v, instead of v^2, or to the product of v^2 and the surface area, which could be decreasing through evaporation. Investigate at least one alternative model.